The Science of Winning

by Danilo Lapegna

How scientific method can help us become smarter, wealthier, happier.

Winning… thanks to science? 1
I - Recreating success in the laboratory 9
Strategy Lab - The SPM 13
How much research? How much study? How much experimentation? 18
The mind is dumb 21
Strategy Lab - The Science of Deception 36
No numbers, no party..39
Strategy Lab - The Mad Numerator 47
II - Mathematical Distributions and Buddha 52
Strategy Lab - 80/20 Questions 56
Strategy Lab - Three Approaches 57
Everything has a limit 58
Strategy Lab - "In and Out of the Box" 59
The Secret of the Super Lazy Ones..60
Strategy Lab - "Follow the Buddha and Then Kill Him" 67
Strategy Lab - Swarm 70
III - The Dopamine Map 74
Actions for the "negative aspects"..77
Actions for the "Positive Aspects" 83
"Universal" Strategic Guidelines 92
Strategy Lab - Costs/Benefits..96
Strategy Lab - Project Pillars..98
Strategy Lab - The Eight Forces 102
Strategy Lab - The Seven "Muda" 107
IV - Scientific Hacking..112
Useful discontinuity..114
Moving "sideways" 116
Predictability 118

Shifting focus..120
Rethink the rules..122
Upper limit..124
Inevitably interconnected systems 127
Dependency..129
The drop hollows the stone. 131
Lack of antibodies..135
Excessive force...137
The human element 139
Strategy Lab - Scientific Self-Defense 142

V - Towards the Core of Truth..145
Start with some assumptions 146
Wild pruning..148
Interactions and receivers..150
Similarity and Past...152
Build the hypothesis model...153
Change and retry...156
Ripples in the water...158
Waiting for Spring..159
Strategy Lab - Scientific Intuition 160

VI - The keys to Mastering the Future 166
How to predict the future 167
The Formula for Success...170
Strategic and tactical variability...177
Playing with fire 179
Strategy Lab - The Science of Winning.....................................188
Strategy or science? 193

Thank you for reading!...198

But there's more... 199
The Author 206
Bibliography and Further Reading 207
Disclaimer 210

This isn't an AI-generated book, and that's by design.

Yes, we use cutting-edge technology to polish our writing and augment our research, but **the heart of this book is all human:** painstaking research, deliberate crafting, and *a lot* of late nights! Just so that you know: **what you've got here has been shaped with heart, intention and care!** Oh, and also, in case you don't like something: you can blame us, not the robots!

Winning... thanks to science?

What exactly is *science*?

Because when faced with the word "science," our mind can start to conjure up a thousand images, from unfathomable subatomic mysteries and indecipherable mathematical formulas to the image of laboratory hermits who are ensconced in their dogmas and speak languages incomprehensible to "ordinary mortals."

And here I could tell you that science is that "wonderful thing" that has allowed us to transcend the limits of our senses, enabling us to explore all the secrets of the infinitely small and infinitely large. It is what has fulfilled our ancestral dream of having wings, even laying the groundwork to set foot on distant planets in remote galaxies. It is what has allowed us to discover the "building blocks" we are made of, making us stronger, more intelligent, and longer-lived. Through the application of technology, it has reduced the hours of our workday, granting us more leisure time and hence the opportunity to pursue our interests, "have more fun," delve deeply into our desires and our innermost nature.

But unfortunately, what has been said so far only skirts around the question without answering it, so let me start over: "science," simply put, *encompasses all knowledge derived from the application of a scientific method.* And what exactly is the scientific method? Admittedly, this isn't an easy question, but let's give it a try: it is a procedure for discovering reality based on a certain rigor and formalization in data collection and problem formulation. Additionally, it is carried out through verifiable, reliable, repeatable procedures and, importantly, procedures that can be shared and, if necessary, *proven false*. All this with the necessity of using *experimentation* as the final validation method.

This is clearly not without dilemmas, due to the complexity of both the method itself and the reality in which we live. However, despite these difficulties, what matters is that the scientific method remains the best possible way to gain a deep understanding of reality; what exactly is meant by "reality" will be examined in more detail later on.

Before science, "we weren't doing great"

If you've read even a single history book, you'll surely know that before the so-called "Scientific Revolution" (16th-17th century AD), knowledge and technological innovation were much more influenced by dogmas and beliefs that were neither verifiable nor refutable. This certainly doesn't mean that communities accepted things as they were or outright rejected any logical or rational approach to the world; however, it was "generally much easier" to rely on dogmatic assertions, or those based on the opinions of the most respected authorities: Aristotle, the Pope, the Church, the Emperor, or the shaman of the time, etc.

And so, despite the obvious historical presence of significant progress in certain areas, as demonstrated by the development, for example, of astronomy, architectural techniques in antiquity, or the craftsmanship arts during the Renaissance, the absence of a scientific method (and the entire community and technological structure underpinning it) has nonetheless set an "upper limit" on our ability to advance. This has allowed for the persistence of evidently harmful practices: consider the repeated treatment of diseases through bloodletting, the lack of adequate hygiene standards even in the most prosperous cities, or the complete justification of forms of violence or abuse in the name of certain social orders.

Without a method like the scientific one to "support" the process of updating knowledge, it was much easier to fall, even for centuries, into traps of all kinds; and it was too often easier to rely on (and remain within) knowledge that not only was arbitrarily constructed but also often very skillfully defended. A big problem in fact arose often when these doctrines collided with tangible evidence; since any criterion for an "appeal to scientificity" or the verification of what is real wasn't necessarily there, there was always an easy escape to silence moments when reality "came knocking." Here lies the inadvertent comedy of dogmatic knowledge, still persistent today in the thoughts of authoritarian ideologies, sects, and individuals in bad faith of every kind: they are often constructed in such a way as to be impossible to verify or refute, or they are defended by denying any possibility or need for confrontation. For instance, if that curative method invented by the Supreme Alchemist Ezekiel in the fourteenth-century citadel of Qualiqquà led to thousands of deaths, it is likely—especially if Ezekiel held particular power in the citadel—that the death toll was considered misleading, erroneous, perhaps the result of some conspiracy (which, once again, "forcefully brings us back" to the present, as such rhetorical subterfuges are the desperate refuge, for instance, of dishonest politicians and "supergurus" of all kinds).

And yes, despite the fact that many scientists today unfortunately think this way, science (and yes, it's crucial from the start to distinguish between these two things: the scientist *is just an individual*, while science *is the communal result of the work of countless individuals*) in the case of evidence proving an error, will always be "sooner or later" forced to admit the mistake and change, expand, update itself. This also immediately reveals why scientific thinking "wins" when it comes to understanding reality, and thus learning how to change it: in its constant evolution in front of new evidence, this method is also "inevitably" what gives us the greatest likelihood of having something that not only *works* but does so *in a lasting, reliable, repeatable manner*. It's no coincidence we can hope to survive diseases that only a few decades ago were 100% fatal because of what science has revealed to us; not thanks to the writings of the aforementioned Ezekiel of Qualiqua, nor any of his colleagues.

Fair objections

I know what many of you might be thinking right now: okay, but science *doesn't provide all the answers!* And yes, you're right; however, some answers, such as those related to aesthetic, moral, or "spiritual" issues, simply do not fall within the domain of science; for certain truths, one must simply look elsewhere. You can't blame the designer of the fork if you use it to eat soup. Or another example: *science makes mistakes!* And yes, it's very true, but be careful: if data is gathered through the scientific method, this doesn't guarantee its "absolute" accuracy. A simple example: if you read academic texts in biology or medicine from two hundred years ago today, it's almost certain they are full of nonsense. However, the key lies exactly in what was mentioned before: in the "intrinsic" ability of the method to *embrace doubt initially, and then error,* to allow knowledge to *update itself* where necessary. Therefore, if, in the face of scientific errors, we were to question the entire scientific method, we would be committing the intellectual equivalent of discarding an entire recipe just because it lacks a bit of salt.

But now, an absolutely legitimate objection: *"Many times throughout history, science has taken decades to admit things it initially denied, which were obvious or apparent."* Yes, exactly, a weakness of science is its inability to provide certainties without sufficient evidence, which, simply put, often causes it to lag behind. However, this is also one of its greatest strengths: its inability to make pronouncements without sufficient evidence tends to ensure the presence of a fantastic "filter" that, in most cases, prevents the acceptance of the delusional theory shouted by the prophet of the moment into scientific knowledge. For everything else, for all the cases in which science cannot provide a definitive or immediate answer, our natural pragmatism and desire for experimentation can always lead us to alternative methods of interpretation and understanding. It is crucial, once again, that this idea of being "more scientific" in our daily lives is part of a global, holistic vision, where, yes, even a "spark" of irrationality, intuition, or "madness" is allowed (where applicable and sensible, and especially where it does not lead to self-destructive nonsense). Because yes, let's emphasize it once again: science can offer us the means to explore only certain well-defined domains of existence; for everything else, we can draw from it some principles of rationality, logic, rigor, and common sense, but it will be impossible, if not undesirable, to remain "scientific."

The love for experimentation and the search for what works

Having made our "detached declaration of love," a "I partially, reasonably love you" to the scientific method, it won't be difficult for us to reach the "conceptual heart" of our text: a core that draws energy from this method's ability to reveal *what works best*, providing us with an extraordinarily valuable set of strategic and tactical tools. Whether it's improving our monthly earnings or academic results, overcoming this or that crisis, losing weight, or finding more time to spend with our families, learning to apply the "spirit" and methods of science can prove to be the best way to gain a clear and expansive view of the reality that concerns us, and thus aim to extract the maximum result in exchange for the minimum effort.

In this book, through chapters of theory, stories of our "Scientific Champions," and the practical mini-sections of the "Strategy Labs," we will try to understand how to fight our battles through research, rationality, application of mathematical and logical principles, continuous experimentation, and the use of errors as resources. Most importantly, we will maintain a strong "natural love" for an honest, objective, pragmatic perspective free from dogmas and ideological traps. This, in my opinion, is absolutely essential for a sensible approach to the world; it becomes even more crucial in times as confusing as ours. We face entirely new challenges and see the resurgence of anti-scientific ideologies, mystical theories about flat earths, and a magical tendency to propose "miraculous cures." Therefore, it's more important than ever to rekindle our "love" for reality, and based on this indispensable "pillar," evolve not only as individuals but as a fundamental and driving critical mass for what can be considered a bright future. A vision that, much like in the sci-fi utopias of the past, knows how to catalyze the grandest and noblest dreams of each of us.

"Science is magic that works."
(Kurt Vonnegut)

At the "Kintsugi Project," we celebrate the joy of reading every day. If you've got a physical copy of this book, share the love! Snap a photo and tag our Instagram account, @danilolapegna.kintsugi, using the hashtag #kintsugibooklove. We'll be delighted to send you a personal thank-you.

For feedback, proposals, requests, or suggestions, don't hesitate to reach out to us at info@kintsugiproject.net. We're always happy to hear from you!

Danilo Lapegna

- The Science of Winning -

"The scientist is free, and must be free to ask any question, to doubt any assertion, to seek every evidence, to correct every error."
(Julius Robert Oppenheimer)

I - Recreating success in the laboratory

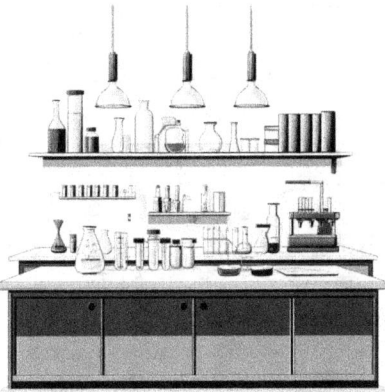

As mentioned in the introduction, if there is one principle that might be trivial for some but not for others, and from which science cannot deviate, it is the existence of reality, or something objective. Or, if we want to present it on an even more "bare-bones" level, it's that element that will behave in a certain way regardless of how much we choose to ignore it. Consider the very simple example of gravity: if we jump off a balcony on the thirtieth floor, well, we will die even if we decide not to believe in it.

"The great thing about science is that it's true whether you believe in it or not."
(Neil DeGrasse Tyson)

And here I already see that many of you might start to exhibit a risk of absolutist and monolithic thinking, where the scientist wants the universe to be solely what he (or she) desires, without room for debate, different perspectives, or free thought.

Yes and no, let's try to calm down and delve into the matter for a moment. First of all, as suggested in the introduction, not everything can be considered objective or real, and we cannot think of applying a scientific method to every existing problem. This has never been our goal nor will it ever be.

If we discuss, for example, any field dominated by the *"hard sciences" such as physics, chemistry, or biology*, a scientific method based on a rigorous collection of data and truths established through repeatable experiments becomes necessarily the only possible way to gain more knowledge. Simply because in these disciplines, the "balcony principle" mentioned earlier applies: every ignored truth "remains there," "exists regardless," whether explored or not. You can debate, not believe, or claim it to be different all you want: gravity, entropy, are there, ready to prepare you for your encounter with the sidewalk should you suddenly believe you can fly by jumping out a window. When we "navigate" through the hard sciences, we can thus speak of objectivity and reality.

But just take a microsecond glance at various fields of knowledge to realize that there are countless contexts where things become vaguer and less defined. Perhaps because the statements that can be made in that field are inherently less verifiable, or because the observable phenomena are much less predictable, demonstrable, and reproducible. Or, if we want to adopt a purely Popperian criterion (from the philosopher Karl Popper), it is no longer possible to apply the criterion of *falsifiability* to the laws involved; that is, the chances of conducting experiments that can objectively refute, or disprove, a statement will dwindle, possibly to zero. In short, you can't objectively prove this thing, nor can you objectively prove that it's not so, and thus the chances of saying "this is real" progressively diminish. Things may become more *probable* or more *agreeable*, but they move out of the realm of what is objectively existent or valid.

Consider, for example, what some call "soft sciences" (for some, not even considered sciences, but for simplicity, we will continue to classify them as such) like economics, psychology, or meteorology: there is a basis of real laws (like chemical or physical laws that govern their phenomena), there are models for reasoning and making predictions, but as you may have noticed every time it started raining on you while you were without an umbrella, these models can vary, decline, fail, or be partially reinterpreted.

However, consider also the types of knowledge that are not technical-scientific at all, like artistic or aesthetic ones, which, as they say, can only be real in an *intersubjective, communal* space of "common agreement and debate." For instance, if you have a copy of the film "The Blues Brothers" in front of you, it's objective, and objectively demonstrable and verifiable, that it was directed by John Landis, that it's from 1980, and that John Belushi starred in it. It's even a fact that it entered the Guinness World Records for the most car crashes in a film. But is it a "masterpiece"? Well, it could certainly be for some. Perhaps for the majority of those who have watched it. It may adhere to some aesthetic principles described in specific film manuals. But here there are no experiments to conduct, and there's no "reality that will take care of you even if you ignore it"; rather, meanings are constantly redefined and redefinable in the debate. So no, no matter how much you might enjoy the film or how much it might be recognized as a masterpiece by a segment of global critics, we are completely outside the realm of science. Therefore, despite the fact that many professional critics (or even some philosophers, since the debate is always open) might disagree here, we are outside the field of objectivity.

However, considering that a statement can be studied and may be probabilistically more valid than others even if it does not "live" in the field of the objective, I think that many principles and methods of scientific analysis, if applied to these fields, can still provide us with reliable data and useful guidelines. This is demonstrated by the fact that we do not only study how to calculate 1 + 1, but also how to excel in completely non-scientific fields such as theater, music, or literature. With some exceptions, foremost among these certain metaphysical beliefs that literally have no verifiable or falsifiable basis, we could say that a good "quasi-scientific procedure for success" can always be structured as follows:

- **Study and research** of the involved sector and its internal aspects.

- **Extracting useful information:** Analyzing potential cause-and-effect relationships or other connections between things, searching for unknowns, studying the evolution of involved factors and their structure. Discarding all non-verifiable or less verifiable information, with a priority focus on what seems to have a more objectively valid nature, and developing one or more hypotheses on what may work for our primary goal.

- **Testing of the hypotheses through action** (or trials and simulations of the same, if feasible), adjustment following possible errors, and re-testing until the desired goal is achieved.

We could call these three steps the perfect summary of a sort of small "scientific-pragmatic method," or SPM: the "three-step guide" to progressively extract some of the rules that are most likely to be useful for one's purpose. This may seem trivial at first glance, but it can be extraordinarily valuable in all those fields of action where we might be "stuck" because we're trapped in misleading truths. In such cases, "forcing ourselves" to study and experiment can help us escape our maze of preconceptions and thus achieve, without unnecessary or harmful filters, a set of laws and information that, by offering to change our perspective, might also reveal new and extraordinarily interesting aspects on which to act.

Because, and we'll explore this further later on, sometimes simply accepting and implementing information never considered before can really change everything. Think of all the countless practical and non-scientific fields where this principle is applied: in the art of war, the use of spies to gather "secret" information from the enemy is considered one of the cornerstones of the entire theory. Or, in the context of negotiation, what the FBI calls "Black swan-based situations," a term inherited from philosopher Nassim Nicholas Taleb who spoke of a "Black Swan" as a particularly rare event that "changes everything." A type of FBI negotiation strategy involves trying to uncover a hidden but particularly important truth for the counterpart being negotiated with (consider a trauma, a fundamental rule of conduct, or a set of rituals they cannot forgo) and using it as a "game changer" to bring them over to your side. This may not be as scientific as Kepler's three laws, but it certainly saves more lives compared to tackling critical situations "by instinct" or "randomly." Some might argue otherwise, "downgrading" anything that is not strictly scientific (like psychology) to the level of "nonsense," similar to astrology. Although I acknowledge the evident limitations of disciplines built on a statistical or empirical basis, I would feel inclined to ask those detractors: *"When you face difficulties of any kind, do you consider a consultation with a psychologist as valid as one with an astrologer?"*

Strategy Lab - The SPM

Now that we have introduced our first possible application of the scientific method to real-world problems, let's try to put it into practice through a small "laboratory exercise" that follows the structure study → extraction → experimentation → adjustment. More specifically:

1 - Choose a problem to work on. Try starting with something simple yet frustrating. The car that reaches fifty-eight degrees when you go to pick it up in the morning, the excessive electricity consumption in your home, the fact that you've started gaining weight too quickly in the last two months, and things like that. Nothing too lofty or life-and-death issues, but something that you simply feel is particularly relevant on a personal level.

2 - Conduct some study and research. Devote a "reasonable amount of time" solely to studying and gathering information related to this problem. An hour, half an hour, two hours... I'm sure you can determine for yourself what's reasonable. And do it by asking yourself, for example:

- How have others managed to solve similar problems?
- What has proven to work better here?
- What generates what?
- What in this situation "weighs 10"?
- And what about "weighs 100"?
- Could I intervene on the causes? On the effects?
- Are there strategies, practices, actions that are certainly effective?
- What are the unknown variables? How could I uncover them?
- Are there relationships between the elements involved? Things that could be connected to each other? Elements that could "unexpectedly" be the cause of effects that interest me?
- Am I missing information that I have never wanted or been able to research? Could I undertake "unconventional" studies and explore "unprecedented" aspects?
- Am I missing a resource that I never wanted or was able to acquire? How could I obtain these things?

A primary piece of advice I can give regarding the "philosophy" for conducting this investigation is to avoid perfectionism and not aim for complete informational coverage on the situation in question. You will clearly understand what works and what doesn't in the next step. Second piece of advice: use this phase to begin "training" your brain a bit in source selection. That means, given two completely contrasting pieces of information on the same issue, where does the reality lie? Which of the consulted sources is more likely to be reliable as a seeker of truth rather than influenced by other factors? (such as ideologies, emotional responses, validation of one's own beliefs or experiences) Is there a truth that, perhaps with varying emphasis, unifies all the information from the different sources? This is a crucial aspect that we will explore in greater depth later on.

3 - Testing the hypotheses. Now try to concentrate primarily on acting based on what you extracted from the previous point and ask yourself:

- Can I start doing something right now?
- Can I put together a plan?
- Is there at least one set of actions I can take based on what I know? Or regardless of what I don't know?
- Can I, regardless of the uncertainties, take action to create something that will still be useful, reusable, and "resellable"?
- Do I have enough time and resources to allow for mistakes and failed attempts?
- Can I venture a hypothesis on what works and refine it through action?
- Is there a real risk in taking action "in the dark," or is it only fear, laziness, habit, and fear of "change" that are speaking?

Define precise criteria by which your problem can be considered solved, in terms of quantities and deadlines. As we will see later, while not everything is measurable, it is true that the scientific habit of working with precise numerical quantities allows for the greatest hold on reality. Therefore, we could formalize our problem not as "losing weight," but as "losing two kilograms within a month." Not "earning more," but "earning at least 20% more within a year."

Finally, if within the given time frame you haven't reached the "quantity" you had set, try investigating what might be wrong with the method itself or the criteria by which you applied it; then change only that, keep the project and the overall vision, and set yourself a new deadline.

This "strategy lab" method, as I'm sure you've already noticed, isn't particularly in-depth or rigorous, but it can be useful for solving small, not very complex problems. Moreover, it can provide excellent mental training to begin adopting those typical "mental states" through which a scientist understands and, consequently, attempts to define the criteria for modifying a reality: a love for deeper knowledge, acceptance of one's interpretative limits, a certain rigor in defining what works and what doesn't, verification of the made hypotheses, and *viewing error not as a breaking point but as an inevitable feature of the process itself.*

Given the fundamental importance of the topic, I find it essential to spend a bit more time on the historical and philosophical aspect of the *error*. An incredibly interesting explanation on the connection between error and science was presented by the American philosopher Charles Sanders Peirce, who, towards the end of the 19th century, spoke of abductive reasoning. This term had been used since Aristotle, but Peirce expanded it, defining its constitution and application as the first true step towards scientific thinking.

Premise: **deduction, induction, and abduction** are three methods to expand one's knowledge. **Deduction** is the process wherein, given that we know for certain all the apples in a basket are red, and given that we will take one out, we can predict that the apple we take will be red. This moves from a general rule to certainty about what will happen in a particular case, with no possibility of error. **Induction** is the exact opposite, meaning the idea that, after continuously taking apples from a basket (perhaps covered by a cloth), and finding they are all red, it is possible to deduce the rule that the apples in the basket might all be red. This, as can plainly be deduced from the conditional phrase just used, is not a logically perfect procedure since it would only take drawing a single yellow apple to invalidate the newly constructed rule.

And then there's **abduction**, similar to induction, with the difference that its aim is not to extract a rule but to form a *hypothesis*—something that is already known not to be certain, and thus needs experimental confirmation. For instance, I see red apples on the ground and a basket in which I know for sure all the apples are red, hence I hypothesize that the apples come from that basket. In the book "The Sign of Three," Umberto Eco provides a very similar example of abduction: "If you see tuna on your plate and an open tuna can on the table, you can bet that you will certainly think the tuna on your plate came from that can, but this is only an abduction." Eco, known among other things for his work on Peirce, further adds that in Doyle's novels, Sherlock Holmes calls "deductions" what are actually "creative abductions."

Given the premises that the character elaborates about the crime scene or the suspects' behavior, there is never such logical consequentiality that allows reaching the culprit with the same mathematical certainty as in the apple example above. Indeed, generally, the more complex the systems being examined, the greater the possibility of hidden variables, making it more challenging to say, "if this is the premise, then this is certainly the consequence."

However, the "genius" of the character lies in formulating abductive hypotheses that can creatively relate even the most challenging elements, and by combining this with his vast knowledge and experience in the field, such hypotheses end up having a very high probability of being true. It's no coincidence that Holmes often makes mistakes, and his investigative process cannot be considered complete until there is experimental verification, in his case an "incontestable" proof that his hypothesis was correct.

On the other hand, I know that despite our objective need to proceed with "repeated abductions," as individuals, we still struggle greatly to accept the concept of "mistakes." Very often, this is because many of us still have, perhaps subconsciously, that "annoying school voice" in our heads that, when faced with an error, continues to threaten us with the image of a "dunce cap." But the "secret" here lies in trying to understand that today, we can reach space and live healthily for many more years precisely because millions of people have worn that cap as a sort of "necessary rite of passage." The moment we learn to silence this presumptuous voice and understand that every mistake, every misstep, is not only inevitable but often the unique resource through which reality reveals new tools for us to grow and build, we can say we have made our "triumphant entry" into a much broader world.

Because where clarity or depth of vision is not granted to us, an experimental approach becomes the only possible answer; and it is precisely thanks to this "knowing how to advance despite setbacks" that, even in the face of uncertainties and obstacles, we can learn to achieve extraordinary things. As one scientist, who was perhaps also the most brilliant scientific communicator of all time, Richard Feynman, said: *"Slowly, it is the very errors themselves that guide us to the truth."*

How much research? How much study? How much experimentation?

Given the previous "strategy lab," a natural question arises: how deeply should we dive into our research? When is the right moment to pause studying and shift to experimentation, hence fully embracing the inherent risks of error?

In fact, while generally increasing one's levels of research, analysis, and information extraction is certainly a good thing, it should also be stated that more research and study do not necessarily always lead to better results. On the contrary, sometimes directly experimenting with partial truths and unverified theories can simply be the best way to achieve our goal.

Never think that in this book we want to portray as virtuous the figure of the fearful person who locks themselves in a "laboratory of analysis" to conduct a thousand unnecessary investigations whenever something needs to be achieved. On the contrary, we will always encourage trying to act with a mindset that balances the investigative aspect with the experimental one; all while inviting you to develop the necessary instinct to understand when the former must be entirely sacrificed in favor of taking active, decisive action, even taking bold risks, if necessary.

But since achieving this ideal balance is not always simple, let's try to identify three specific factors that can serve as indicators to help us better understand how much to rely on "intuition, faith, and improvisation," and how much to pause and think in order to research, analyze, and "strategize" before taking any concrete steps:

- **The risk of acting in ignorance:** As both the likelihood and severity of potential dangers in a situation rise, thorough preliminary research might be necessary to minimize damage and maximize gains and benefits. However, when risks are few, unlikely, or perhaps only psychological, you can aim for a phase of more intense action and experimentation. Pay attention to how we mentioned "only psychological" risks. What was said earlier about the impact of personal insecurities contains the seed for one of the most frequent and significant errors in our ability to perceive reality: confusing *actual risks with those just perceived as such*. And this is a topic we will delve into shortly.

- **The "studyability" of the field in question:** Is the field governed by dynamics that can be modeled through the laws of a hard science? Are there clear, precise, and stable rules and patterns? Is it possible to uncover its unknowns? Or is everything much more undefined?
  The more the factors involved show regularity, patterns, and *reproducible cause-effect relationships,* the more comprehensively the field can be studied. Conversely, the more uncertainty, instability, or absence of laws there is, the more likely it will be necessary to focus on action (possibly hoping that it might lead us also to extract new useful information along the way).
  And this idea of "moving forward in cases of non-studyability" can be an excellent source of learning whenever one finds oneself in unclear practical situations: take one step at a time, learn what you can, and initially try to rely on the most universal tools and principles; those with the highest probability of validity, *regardless* of the type of situation or uncertainty faced.

- **The time at your disposal:** Simply put, when we have sufficient time to solve our problem or achieve our goal, it might be advantageous to increase our time for "laboratory and library," thereby becoming more efficient and effective during the action-experimentation phase. However, when time is scarce, it's better to become "bolder abductive thinkers," scientists who first hypothesize, then experiment with partial ignorance, and finally try to obtain confirmations and new information directly during the process.

For example, the not-so-scientific "game of seduction" that many people engage in, often ineffectively, to win someone's attention is typically based on:

- **Low predictability in preliminary study.** Let's consider the case of someone about whom nothing can be known in advance. Even if this weren't the case, little would change, as it's difficult to derive reliable laws about human behavior in most cases.
- **Limited time:** Again, this is not a universal rule, but let's assume for the sake of this example that if you spend too much time trying to win someone over, you might end up "losing" that person.
- **Low risk.** Because even if putting yourself out there and risking rejection triggers various psychological alarms, and can make you very upset, a rejection is not the end of the world.

And it is precisely for this reason that, when one tries to win over someone's heart and attention, it can easily become a game of "risky assumptions": you always start with what is most likely to work in general cases; then everything useful that can be used to improve the effectiveness of one's efforts, such as the other person's tastes, preferences, and trends, is only deduced in the course of action.

On the opposite end of the spectrum, for instance, we have the way medical research is conducted: medicine stems from "hard" sciences like chemistry and biology, and the related research is based on medium-to-high time factors (it's clear that each research project has its own distinct time limits, but let's take this simplification for what it is) and high risk (errors in drug composition can lead to disastrous mistakes).

And that's why, although every science requires a phase of hypothesis formulation and experimentation (which inevitably involves some level of guessing), it is a field where one cannot afford for these hypotheses to be *completely crazy*. One must experiment cautiously, starting from abductions based on rigorous observations or well-validated knowledge. Then, in the worst-case scenario where the experiment is unproductive, one will still reanalyze what went wrong and try to understand how to leverage that knowledge; or perhaps, why not, toward other objectives, just as has occurred countless times when, for instance, a vast amount of research has been contributed to medicine stemming from the field of space exploration.

The final principle worth mentioning in this context: paying attention to what has been said so far about "Laboratory vs. Action" becomes crucial when facing too many unknowns, our insecurities trigger new alarms, and suddenly the laboratory, the "library," becomes a refuge to avoid facing risks or uncertainties. We'll explore this topic in more detail later, but for now, consider those times when you instinctively want to "hide in your laboratory"; it might be that you're simply looking for a way to shield yourself from a reality you don't like, and that improving it would just require an active step, however hard or risky it might be, in the most obvious direction.

"Never give up on work, on commitment. This is what gives you sense and meaning; existence would be empty without all of this."
(Stephen Hawking)

The mind is dumb

The "primary tool" at our disposal for extracting information from reality is, quite simply, the use of the brain and the senses: since ancient times, we have tried to understand the world by touching, smelling, seeing… and by doing so, we achieved both fantastic technological revolutions, like those that allowed us to start cultivating land and working metals, and also a lot of spectacularly foolish ideas, like believing that the sun revolved around the earth and other similar notions.

Due to our biological constitution, we tend to *trust* what we experience and to attribute some sense of rationality and reliability to the way we process these experiences.

Obviously, in most cases, we have to trust what we perceive or intuit, otherwise we would go insane. If, when we press the "E" key on our keyboard, we started to think that not only might the letter "E" never appear, but perhaps the computer would Explode or trigger an Extradimensional invasion, we wouldn't even be able to write a ten-character email address. But even just the fact that historically, by trusting what we perceive, we have managed to emerge from caves and engineer our first technologies is a sign that often what we perceive is more than sufficient for our practical purposes.

However, as the very history of scientific progress shows, there are phenomena that, in order to be understood and mastered, require a much more complex and extensive study than what our senses, our brain, and the most readily available tools suggest. Just think of how many centuries it took us to fully understand the nature of the structure and movement of the "luminous circles that moved in the sky."

And that's why, as good scientist-strategists, whenever complexity increases, we must embrace the concept of "Socratic ignorance," thereby committing ourselves to face *reality;* this, with the understanding that our perception is *partial* and that there is likely more than what we see, hear, and contemplate. More importantly, we need to strive to become aware that, as we perceive and process them, aspects of our experience might have been distorted by simplifications, alterations, and cognitive or perceptual errors.

This, mind you, should never devolve into doubting only what doesn't conform to our beliefs and then chasing what validates them (consider those who reject every official explanation to always seek a conspiracy or hidden truth). It should instead focus on having the courage to let go of the notion of completeness or perfection in one's convictions or perceptions, embracing the possibility that there might be more (or something entirely different) beyond what we currently know or believe.

A particularly interesting aspect of this discussion is that when it comes to assessing *quantities,* our brain often performs poorly. Quantity is a fundamental element in any complex discourse, which is why we've historically started adopting precise tools for rigorous measurement. For example, using terminology from the latest "strategy lab," and which will be crucial later on, we too often believe we can easily distinguish between an element that "weighs 100" and one that "weighs 10." Or worse, we completely *forget* the concept of quantity (how precisely does this thing "weigh"? how does it weigh compared to another?) in favor of simplifications based on "yes/no" or "black and white."

And so, for example, if during our journey we encounter an obstacle that might be quite simple to overcome, yet the thought of it provokes in us an enormous fear of the conflicts we will have to face, we might believe we are facing a "10,000-pound obstacle." As a result, we might even end up abandoning a path where, perhaps, all it would have taken to resolve things was a half day of dedicated work.

But this is a topic often analyzed in studies on the impact of media communication on populations: many of these studies tend to agree that countries where journalism and politics tend to have more emotional and "sensationalist" communication are typically correlated with a population much more "ignorant" about how each issue actually impacts the society in which they live. This is precisely what certain "populisms" are based on: the politician who shouts from their social media page that this problem is "ruining the country," using tones, content, and stories with easy emotional appeal, also easily leads their "followers" to believe that this problem actually has a or "10,000" impact; all while perhaps the numbers, statistics, and concrete impact on public spending are completely negligible compared to other facts that might be overlooked because they are not easy "fuel" for the dynamics of propaganda.

Having made this lengthy but necessary introduction, how can we minimize the impact of these cognitive and perceptual errors and thus achieve an "improved" ability to understand reality?

The first step has already been mentioned, and it is as simple to state as it is complex to implement: it is based on the principle that by gaining greater awareness of what influences us, we could gradually "dismantle" its power. You can begin by paying more attention to yourself and the moments when you allow the most "suggestive" emotions or images to take over. By stopping in front of these moments and trying to tell yourself: "Hey, this is just an emotion," "Hey, this is just a suggestive image," "Let's try to go deeper than this." The complexity of this process lies in the fact that, in most cases, we are so "gravitated" towards our thoughts and emotions that they end up representing the very universe in which we live. An analysis of how to increase our detachment from them would require a separate book. However, for now, start by more frequently adopting the principle that your thoughts and emotions *are not the universe*, but are simply "just" thoughts and "just" emotions. Structures that, although they obviously represent the most evocative and engaging form of experience for you (after all, they are your individual and constant "virtual reality"), remain partial and fallible; and as such, will greatly misguide you on the actual impact of things or events surrounding you. Think, for example, of all the times you felt a terrible fear for a challenge that turned out to be a walk in the park. The more we realize how "dumb" we've been in the past, the less dumb we'll perhaps be in the future.

"All that can be learned from a lifetime of science is the vastness of one's own ignorance."
(David Eagleman)

A second step, which reinforces the first, involves gaining a deeper awareness of the actual *biases* that drive us. If we want to use the more specific definition by MG Haselton in "The Handbook of Evolutionary Psychology," biases are *patterns of systematic deviation in one's judgment.*

The history of scientific research into the identification and classification of biases likely begins in 1972, when Amos Tversky and Daniel Kahneman began to analyze in the lab the human inability to rationally evaluate very large quantities. From this issue, the two professors and their collaborators delved into the "errors" and blind spots in the process of human decision-making and cognition; the various "shortcuts" that, often for the necessity of saving time and energy, are continually preferred over a "slow and costly" process of analytical or rational investigation. A particularly famous case was the "Linda problem": a group of people was given a description of an imaginary person, specifically a certain "Linda," who in the example was supposed to be "politically engaged and very concerned with social issues." Then, participants were asked which scenario would be more probable between the following cases:

1. Linda is a shop assistant.
2. Linda is a militant feminist.
3. Linda is a shop assistant *and* an active feminist.

Before I tell you the solution, and if you don't already know the answer to the question, try to answer it yourself. Done? Which of the three is more likely?

The answer is... none of the three! That's right, there is no objective, verifiable, and reliable criterion that suggests the premise leads to one of the three answers with a greater probability than the others. Whatever your choice was (if you chose one of the three and didn't answer, as you should have, that "there's no evidence to support any of them"), what guided you was, indeed, a bias. You made a "cognitive leap" and both your experiences and mental shortcuts suggested a "quick" probable solution to reach a conclusion. This is in fact, if we want to delve deeper, a classic example of *"Conjunction fallacy,"* where we tend to consider a coincidence of multiple events (convict AND activist) more likely than the individual events taken separately (and therefore "inevitably" more likely from a mathematical perspective). But the opposite is obviously true: as we'll also see more in detail later, if an event A has a probability of occurring, and an event B has its own probability, *the likelihood of both A and B happening together can never exceed the probability of either event happening alone.* Maybe a bit counterintuitive, a great reminder that our intuition often works against mathematical logic.

From all this, it's not difficult to deduce that, depending on our experiences, knowledge, and habits, we will inevitably tend to make certain mistakes more often than others. Therefore, in order to gain greater awareness of all of this, let's try to explore together a list of biases universally recognized as such. The list that follows, especially considering that some research is still ongoing, is by no means exhaustive but can nonetheless serve as an excellent starting point to begin understanding the myriad ways in which we "think wrong":

"A bug is a bug" - Effort Justification Bias

Our mind tends to place greater value on everything that requires, or has required, more effort to obtain. This often propels us towards unsustainable costs or unnecessary challenges just to experience the thrill of knowing we can overcome a wall or barrier (ever spent months longing for something too expensive, only to buy it and quickly tire of it?). However, this distortion also inevitably leads us to overestimate the quality of any work where we have personally invested time and resources. As Antoine de Saint-Exupery once said, it is probably "the time you have wasted on your rose that makes your rose so important," but in the end, other criteria will determine whether this same "rose," for example, will be a commercial success or not. Or as an Italian song says: *"Every bug is beautiful to its mother,"* meaning everything appears beautiful in the eyes of the one who created it. While this may not be a reliable law concerning insects, it perfectly describes a factor of our behavior that we must heed whenever we aim to form a more objective view of reality.

"We are more social than we believe" - Social Proof Bias

Our social nature, genetically inclined both to seek approval from the community and to learn through imitation, naturally leads us to consider something as more or less significant whenever it is presented to us as such by society. A note of caution for all the "rebels" out there: even those among us who are most inclined to challenge the prevailing norms of their culture can fall into this trap; in this case, we simply end up attempting to adopt, and instinctively imitate, the social reference of the rebel with whom we most identify. Therefore, please, never believe that you are too "immune" to this effect; instead, you should always strive to pay more attention to cases in which your judgment is such because its content is socially inherited. Because yes, this will happen much more often than you would like to admit. And it's likely to be your "primary source" of colossal mistakes, whether in evaluation, planning, or action.

"Looking far ahead is essential" - Proximity bias

In other words, if we perceive an element as closer to us, whether in space or time, we are very likely to also consider it more relevant. This especially happens in the evaluation of risks and dangers: we feel much more at risk if a major plane crash or a large terrorist attack has occurred, or has recently been shouted about by the media, rather than if we haven't heard about it for a while. This bias can have significant implications for our daily life and for society in general. It can influence political decisions, marketing strategies, consumer behaviors, and personal life choices. For example, politicians can exploit the proximity bias to influence public opinion by focusing attention on recent and local events or issues, instead of addressing structural or long-term problems. Therefore, in my opinion, this bias should be countered by favoring a comprehensive and extended view of what interests us, rather than what is "fresher" in our memory. Aiming for an overall perspective, a fair assessment of the weight and impact of historical phenomena, is often all we need to escape the trap of "recent noise" and propel ourselves towards a much more objective and well-considered evaluation system.

"It could all be much more complex than that" - Schema bias

The "schema bias" refers to the fact that our mind, often in need of structure and recognizable patterns, tends to simplify the elements it processes by categorizing them into schemas or patterns. This often occurs on multiple levels: on the "purely" sensory level, for example, the phenomenon is known as *pareidolia*, and it's what makes us believe we see animals or faces in the clouds, or even hidden messages in the lyrics of a song.

Needless to say, our mind requires patterns and categories to interpret the world; without these constructs, we could not come to any kind of complex knowledge. Imagine finding yourself in front of a wooden table and, in trying to figure out what material it's made of, you realize you can't even conceive what a material is, then confusing it with a cat, which you still couldn't distinguish from a pair of wireless headphones. But then, what are wireless headphones? This is thought deprived of categories. Now imagine having a book in front of you, wanting to read it, but not knowing which page to start from just because the title is different from the book you read yesterday. This is thought without the ability to resort to pre-learned patterns.

The "problem," however, in the opposite sense, arises when this "reduction into schemes and categories" represents an excessive simplification or a complete misinterpretation of the phenomenon we are analyzing. Consider, for example, what generates a pseudo-superstitious belief like "Murphy's Law," which states that "anything that can go wrong will go wrong." Despite the obvious unscientific nature of such a statement (assuming it is always possible to define "wrong" as the most "destructive" outcome among possible options, why was the universe created then? Why life? Why our birth?), many truly believe it, especially when applying this schema becomes a perfect excuse to avoid uncertainty. But even the story that follows, although perhaps "extreme" as it is based on an explicit intent to deceive, is still exemplary:

"It was the year 755 when in China, during the Tang dynasty, military governor An Lushan rebelled against imperial power. As a result, he assembled a well-trained army of 40,000 men under the command of General Linghu Chao and sent them to besiege the city of Yongqiu, which was poorly defended by a small contingent of 2,000 men led by General Zhang Xun.
Despite the numerical disadvantage, Zhang Xun devised a clever plan: he had his men create a thousand dummies dressed exactly like soldiers, lowered them over the walls during the night, and tricked the rebel army into attacking them with thousands of arrows. The attackers felt deeply humiliated when they

realized they had wasted arrows on dummies, and this feeling persisted when Zhang Xun repeated the dummy tactic the next night.

On the third night, however, the general defending Yongqiu lowered not dummies, but a contingent of his best men. Due to a 'pattern bias,' these men were mistaken for dummies and were ignored, allowing them to go unnoticed as they raided Linghu Chao's camp, setting fire to the tents, killing several men in their sleep, and causing chaos that forced Linghu Chao to flee."

(Chinese Military Anecdote)

Should we then completely renounce our simplifications and structures? Obviously not. But when should we try to "break" them, question them, or expand them with new information? The primary answers here would clearly be: *"In all cases where a framework is unnecessary"* and *"In all cases where the evidence of reality compels us to reconsider the preexisting framework."* This, of course, would cover instances where we might be oblivious and tend to ignore these realities, as in the case of Murphy's Law, but it would not cover the scenario of military deception mentioned earlier, where evidence is deliberately altered. Therefore, I would add a significant: *"and in all remaining cases where it is not too costly to do so."* This, despite its apparent simplicity, actually encourages us to maintain an ongoing openness in our mindset, adopting a perspective that "doubts a little more than necessary," so as not to miss out on the "essential" complexity and diversity of the world.

"Maybe it's time to pull our heads out of the sand" - Confirmation and Ostrich Bias

These two biases, which often operate "in tandem" in our minds, lead us to observe, remember, and give more weight to information that confirms what we already know or believe (confirmation), while simultaneously discarding everything that contradicts it (in other words, we often tend to "bury our heads in the sand," as ostriches are said to do).

This bias, parallel to the previous one, reminds us of the importance of maintaining a certain *openness and flexibility* in our views; but even more, it invites us to consider that often the most valuable information is precisely that which would unsettle us, surprise us, or which we would normally discard because it is "too difficult" to accept. Since we are generally "blind" to them, when we even "hit ourselves" with challenging things to digest, we are much more likely to glimpse a new perspective, a surprising voice, a piece of information capable of changing everything.

But as mentioned a few pages ago, the "confirmation" and "ostrich" biases are also the reason why certain forms of "conspiracy thinking" dominate in some social circles. "Conspiracies" and hidden truths have existed since the dawn of time, and it would be foolish to try to assert otherwise. However, interpreting most of what captures our attention as a "conspiracy," a truth manipulated by someone, is nothing but blindly pursuing certain beliefs, often dangerously disguised as "analytical cleverness." Just like ostriches, those who tend to act this way believe they are free thinkers when, in fact, they are merely slaves to their own beliefs and their desire to dive into some exciting fantasy. After all, what is more captivating? Believing that the real data corresponds to what's shared by the expert, or imagining that the same expert is lying because he is the leader of a cult that gathers every Friday night in a desecrated church to attend satanic masses? And what if we are the only ones who have figured it out, thus representing a sort of "intellectual elite" capable of seeing further than others? Now we're suddenly the main character of our movie, aren't we?

On the other hand yes, we can adopt a more rational approach to a real possibility of hidden truth, or a "conspiracy" if you will: we could start by analyzing the reliability, through cross-referencing, of certain sources and information. We can try to identify logical inconsistencies or lack of verifiable evidence, and only if something's objectively off it would be reasonable to suspect that something is being hidden, or at least misrepresented; all while remaining *ready to give up if irrefutable evidence tells us our imagination was off the mark*. And yeah, the latter is probably what makes the biggest difference between scientific doubt and silly conspirationism. Conspiracy theorists will unlikely give up on their theories, even in front of the most blatant evidence they're wrong. They will just keep their head in the sand, like an ostrich would do. But we can be much better than that, don't we?

"Change will always be frightening" - "Fear of Loss" bias

Simply put: we tend to consider the fear of a loss with much greater emotional weight than a possibility of profit. This is why, in most cases, our brain doesn't care if a change is of quality, because when faced with the prospect of it, the fear of what we may lose will be our first worry. In short, this bias tends to amplify the perception of risks and dangers, often keeping us within what we call the "comfort zone". This bias, and the fear of loss that results from it, is also what leads us to overestimate the value of anything we feel emotionally attached to. So a first great suggestion we could extract from this is: always try to estimate the value of elements, choices, people, and relationships in your life *regardless of how common or familiar they are to you.* At least sometimes *be cold, be objective, be impartial* in your judgement; do it even if you care a lot! This won't just improve the quality of your decisions but will also enable you to take, from time to time, those "leaps of faith" that every life journey often requires; because yeah, sometimes we must learn to *detach,* if we want to *grow.*

The nature underlying this bias also, in my opinion, conveys one of the most precious truths we can ever hold: *"It is normal to fear the future and change* because we are biologically programmed to fear losing what we have; and for this very reason, *future won't be a terrifying amalgamation of all our worst fears just because we believe so."*

This principle urges us, more than anything else, to adopt a "rational trust" in what lies ahead; negative things and losses will unfortunately arrive. However, it is likely that if we know how to "bet appropriately," they will come along with the right dose of positive surprises. Moreover, all the things we daily worry about probably won't happen, or at least not in the way and timing we imagined. From this perspective, the presence of this bias tells us that in the end, "worrying excessively" is not worth it because it won't help control uncertainty; hence, maybe, it's better to "enjoy the journey a little more." So important, so obvious, but still, so often forgotten!

"We cannot be rational where it hurts the most" - Fragility Bias

Everyone has their own vulnerabilities. Each person feels less equipped to handle certain types of situations compared to others, and feels more or less at risk depending on their perception of their ability to deal with them. If you ask around, for instance, there's nothing more terrifying than speaking in public, while for others, the thought of taking a plane is enough to make their blood run cold. Our own fragilities, our own insecurities, are the most relentless "amplifiers of destructive power" out there. As mentioned a few pages back, small, simple things, "with only power level 1," which might easily be resolved with half a day's work, can mentally morph into "things with power level 1000" simply because they touched the "wrong nerve". Now, overcoming certain insecurities is also a subject that would require a separate book, but much can be done when we become more aware of our vulnerabilities and the fact that we will end up overestimating the danger of things that specifically "target" them. This awareness should help us, at least in part, to "recalibrate" our perceptions and improve the objectivity of our perspective.

"Value Can Be Anywhere" - The Appeal to Novelty and Tradition Bias

These two biases, opposite yet complementary, operate differently depending on the attitude and characteristics of the person being considered. Those who are disillusioned by newness will tend to be more captivated by the allure of tradition. Conversely, those "tired of tradition" will be more drawn to what represents novelty. In essence, the idea is that our brain tends to assess something as either less or more "powerful" based solely on its degree of originality or, conversely, its classic nature. This must necessarily be overcome by "bringing back" the evaluation of something to criteria of objective utility rather than to the appeal of its "temporal placement".

"Pirates and Global Warming" - Confusion Between Correlation and Causation

This bias is a more specific variant of the "schema bias" and is a mistake that should be carefully considered in all analytical processes aimed at identifying the causes and laws underlying a phenomenon. This bias is often described by the expression: *"correlation does not imply causation,"* and can be explained as follows: our mind tends to interpret two correlated phenomena (perhaps because they occurred one after the other or repeatedly occur simultaneously) as if *one actually causes the other*. This can lead us to a lot of "interpretation horrors" on real, properly collected data: for instance, in 2005, R. Henderson, a student at the University of Oregon, in the irreverent spirit that led him to start the cult of "Pastafarianism," published an open letter claiming to be able to "incontrovertibly" prove that to combat global warming, the pirate population in the Mediterranean should be restored.

He drafted a Cartesian graph where he could show that, indeed, as Mediterranean piracy decreased, average global temperatures increased by several degrees. The data was obviously correct, but the error *was in the interpretation being given;* that is, precisely an objective case of *correlation* between the two phenomena (which can clearly be summarized by both being caused by human technological and social advancement) was being misrepresented as *causation*.

Obviously, this example is glaring in its absurdity, but there are others, far more insidious, that can easily become part of our everyday lives. Consider all the times when we mistakenly identify something we did as the cause of our feeling better or worse. Or imagine a situation that cannot evolve as we would like. Perhaps we are unable to improve our relational and emotional life or to surround ourselves with people who are meaningful to us.

We might end up thinking that a physical feature we dislike, such as the shape of our nose or the tone of our voice, is the "cause" of the distancing of those around us. After all, both "are there" whenever we present that request for more affection that doesn't materialize, right? Yet, if we looked at reality, we could easily see that such factors are surely not real obstacles to forming relationships. This realization could begin to "dismantle" the original idea, opening up space for a broader investigation. Perhaps if we stop fixating on our nose, we might understand that we've always approached others without any genuine openness and curiosity toward them, which indeed has driven them away.

An excellent way to "overcome" this tendency to confuse the two "c's," ultimately, is to avoid jumping to hasty conclusions, to not accept as valid the cause that seems most obvious, most vocal, or most in line with what we already know or expect (after all, we just discussed the "confirmation" and "ostrich" biases, didn't we?), and instead harness all the curiosity needed to ask ourselves if "perhaps the real causes *lie elsewhere.*"

"Rediscover the Power of the Ordinary" - Allure of Diversity

The episode, the element, the factor that stands out the most for its characteristics and differences from the rest tends to be considered more influential. This bias continuously alters our evaluative abilities. For example, it significantly "enhances" certain events in our memory simply because they differ from the "average" (the one time we were rejected weighs much more than the thousands of other times we felt accepted). Or, it makes us believe that the element with more narrative appeal or originality may be more powerful than what is instead everyday and ordinary.

Imposing oneself not to "fall too much prey" to what appears to stand out from the rest becomes an additional principle of good analysis for the scientist-strategist who wants to better understand the world; the principle of ideal common sense here, in fact, remains that of anchoring oneself to taking as good what works best and vice versa. As banal or ordinary as this principle may sound, it should now be clear to us how much this is the "fundamental law" we tend to forget or set aside most often.

"Does absolute value exist?" - Comparison Bias

In simple terms, the value of something is amplified or diminished when compared to something of more or less value. An ultra-expensive phone seems cheaper when compared to an even pricier model. A four-month prison sentence seems more acceptable when your lawyer presents it as an alternative to four years (and this is an example I hope you encounter only in TV series). And so on.

This requires the challenging effort, for every scientist-strategist, to *extract* certain elements from the contexts in which they are found, or to try to understand *if they would work the same way in a different context;* or, maybe, to ask ourselves how we would evaluate them *if placed alongside something completely different.* The goal here is not to eliminate any possible comparison (after all, we will always need some form of benchmark to make our judgment), but simply to bring it back to a sensible reference within the chosen context.

If, for example, you're trying to determine whether you've met your economic well-being targets this year but then you look at the tax returns of the ten richest individuals in the world, it's obvious that all you'll feel is endless frustration. However, if you examine your tax return from the previous year and notice that you've earned a bit more this year, you've likely adopted a sensible benchmark, something that leads you to congratulate yourself and aim for further growth. Similarly, if you're negotiating a final price for buying a house and a "galactic" number is initially thrown at you, followed by a smaller yet still disproportionate figure, you might instinctively be inclined to accept it. However, if you take a moment to compare the second price (which is seemingly more appealing) to the average price per square meter for the area you've chosen, as well as your actual budget, you might develop a much more reasonable course of action.

Yes, there are many biases and, overall, a lot of information to keep in mind, isn't there? And considering that this isn't an exhaustive list, even within the context of the book itself (as we will introduce others later on), I won't hide from you that it's actually quite complicated to establish a precise method to try and "remove" them from one's thought patterns; moreover, this is almost impossible and could probably also deprive you of some "vital" traits that characterize your personality.

However, my hope is that this list becomes your "go-to guide" to consult whenever you are planning an action strategy: go through the list, try to identify any "mistakes" you tend to make more frequently than others, and see if trying to *adjust* your approach gives you a broader and more constructive view of the reality that interests you.

Repeat this little exercise at the beginning of each work phase and, with time and practice, your "renewed" strategic and analytical abilities will gradually transform into your very own "scientific superpower" to understand the world. This will happen because you will learn to *recognize true relevance,* allowing you to more easily avoid deceptions, artifices, and manipulations; but also because you will hopefully learn to adopt that natural optimism typical of the great "artisans" of science. A kind of optimism that doesn't rely on naive hopes for random blessings, but rather on a grounded positivity born from the awareness that, despite the influence of our biases, the world is full of true potential; a potential ready to be harnessed the moment we stop wandering in the fog of our mind and decide to take just one more step, to accept one more truth, to attempt one more experiment.

Strategy Lab - The Science of Deception

Some of the examples illustrated so far should have led you to a greater awareness of the fact that many of the "biases" of our mind can be consciously exploited to alter our perception. This happens continuously in a world of hyper-persuasive communication like ours: the contrast bias can be exploited to lead us to pay more for an item, the loss bias to make us overestimate the fear of change, the schema bias to create false frameworks to follow, and so on.

Let's hence take a moment to delve further into the topic of "deception" to understand if science can help us learn how to avoid being easily tricked by others; this will also take on a particular significance when we discuss game theory.

According to some studies, the majority of the reasons behind our susceptibility to deception can be attributed to the so-called "inattentional blindness" bias. David McRaney, author of the book "You Are Not So Smart," explained that this bias means we are blind to everything we don't pay attention to, yet we remain completely unaware of this blindness. According to McRaney, we are capable of focusing on only a narrow cone of information at any given time. When we reflect on our experiences, we tend to "fill in the gaps" with a set of concepts "artificially constructed" from what we mentally have available. As a result, not only do we engage in this "informational filling" using the irrationality of countless other prejudices, stereotypes, and biases, but we also do it completely unconsciously. Thus, we end up too often believing that our decision-making and perception processes were much more rational than they actually were, even going so far as to deny evidence just to find confirmation. For example:

- As long as an appearance aligns with pre-existing emotions or past expectations, or at least withstands the amount of effort our mind is willing to exert to doubt it, then this appearance can automatically become "real." Simply put, this is why we are easily fooled by fake news that confirms our beliefs and why we are even less inclined to doubt it if it comes from a website that is skillfully designed to resemble a national newspaper stylistically. A particularly "magnetic" thought, fueled by intense emotion or urgency, narrows our "attention cone," increasing our "ignorance zone" and thus heightening the possibility of being influenced by prejudices, stereotypes, biases, and other misleading information. This is why it is possible that a manipulative merchant who does not want multiple options to be considered will strive to push us to make a purchase as quickly as possible. Or, it is the reason why the art of magic often does not rely solely on the execution of the trick, but on a vast amount of pure theatrical elements, seemingly ancillary but actually fundamental: in fact, if our attention is entirely captivated by the spectacle of these elements, it becomes even less likely that the mind will have "residual energy" to focus on uncovering the ongoing trick.

- If we are provided with information that is "deliberately" sparse or incomplete, even worse if it's "surrounded" by hints that allude to our inner confirmations, we might easily construct truths to complete them based on those expectations, even if on entirely imaginary grounds. This is why we are so drawn to allusions, insinuations, and omissions: for example, when someone responds with unusually brief messages to our "how are you" on WhatsApp, it immediately triggers an alarm in us because the informational incompleteness perceived by us might be quickly "filled" with our fears about their health or our relationship. It is not by chance that this mechanism is often consciously used to arouse attention and interest by those who might perceive they cannot receive it in other ways. This is also why many article headlines try to attract clicks with vague words like "Alert" or "Controversy": simplifying certain facts with such mundane yet evocative words easily suggests to our minds the idea that an "alert has been issued" or that "everyone is engrossed in controversy," even when this is likely not true at all.

- If we are provided with a fully developed conceptual setup, we probably won't feel the "need" to intuit anything else. This principle is the basis of the so-called "double deception," which is the art of showing an initial deception, explicitly leading to its unveiling, and then devising a new fiction through this same revelation. This is the case, for example, of the dishonest person who confesses to having lied in the past: their portrayal as a "repentant liar," as someone whose double-dealing has been revealed, might entirely exclude from our minds the idea that there is a secondary motive behind their repentance. It is also the case of the so-called "Empty City Stratagem" mentioned in the Chinese military strategy manual "The Thirty-Six Stratagems": an unguarded and defenseless city is shown to the opponent → they are led to believe that leaving it this way is a tactic → sensing that an army might be hidden elsewhere, they will abandon the attack.

- If we are presented with "too much" information—more than we can or are willing to process—it becomes highly likely that we will consciously or unconsciously simplify and ignore some of it. This can be particularly deceptive when the other party deliberately manipulates us into overlooking specific details. It's akin to a contract filled with a million seemingly redundant clauses, designed to exhaust our attention, while a crucially deceptive clause is buried within. Overwhelmed by the information overload, we sign it without full awareness; and ultimately bear the consequences.

- Things are not only hidden where no one would think to look, but also, and especially, in plain sight, where nothing unusual is expected; this is because it's highly likely that anything not encompassed by one's beliefs, habits, or prejudices will be automatically overlooked. In this regard, an interesting experiment was conducted in 2014 by Ira Hyman, a psychology professor at Western Washington University. Hyman and his team hung dollar bills on a tree and over two weeks observed that not only did no one stop to collect them, but less than 6% even noticed their presence. This, once again, might lead us to ponder how many resources, how much wealth, how many extraordinary things are hidden precisely in the places we are not accustomed to looking.

No numbers, no party

But which bias among all biases constantly keeps us away from rational assessments and therefore, often unexpectedly, leads us toward making *dreadful* decisions?

A clue: we partly saw this when we started talking about Kahneman's studies, it was hinted at several times a few pages ago and... exactly, it's our innate inability to *work with numbers and quantities!* In any situation that requires our strategic intervention, the most pragmatically valid question to ask should always be: *"How* much does this thing impact?", *"How much good* can it do us?" *"How much* does it harm us?." Yet, except for the rare contexts that require us to measure quantities, we too often decide to base our decisions on purely intuitive, emotional, and qualitative criteria, inevitably leading to terribly approximate and inconsistent interpretations of reality. Consider the classic example of car vs. airplane: the former is hundreds of times more lethally risky than the latter, yet the more "anxious" among us will likely tend to avoid flying rather than avoid exceeding speed limits with their new car.

This chapter will aim to emphasize the importance of precise quantitative assessments for gaining a competitive edge. Whereas in fact emotional-qualitative assessments are often susceptible to cognitive distortions, numbers, when properly collected and interpreted, do not possess this "weakness." They allow us to immediately grasp the real "magnitude" of something based on what we consider significant. They enable easy comparisons between similar entities (does this really matter more or less than that?) and allow considerations that are as "objective" as possible, precise, and focused on what truly matters within the goal we have set.

Obviously, sometimes the quantities we are interested in will be immediately "countable," as in the case of the maximum budget to spend on a project, or the hours spent wasting life on social networks. But in many other cases, we might find ourselves dealing with elements whose nature is particularly difficult, if not practically impossible, to quantify: the affection our partner has for us, the final probability that our project will succeed, the aesthetic quality of this film or that book. If there were objective, quantitative, and infallible criteria to evaluate all these things, we would probably be able to make almost perfect decisions in any situation; but then again, many more things would become completely predetermined, and life itself would transform into something boring and predictable to the point of becoming unsustainable. Which, in my opinion, ties back to what was said in the introduction: *science should not be taken as a "universal key" for all problems concerning the human experience.* Data collection and probability analysis cannot absolutely define the ways in which we approach every aspect of our lives; otherwise, the only guaranteed outcome will be creating problems far bigger than those we might be trying to solve.

However, assuming that we find ourselves in one of those situations where a more strictly quantitative approach can help us put together more accurate models of reality, let's try to look at some guidelines to improve our "quantitative assessments" of the world around us:

Adopt a strategically relevant unit of measurement

In short: try to remember that, often, the only thing that matters is *the unit of measurement* (or units of measurement) *in relation to your goal.*

What does it mean exactly? Let's take a step back for a moment, starting from the most basic premise in the world: every measurement result consists of (at least) *a quantity and a unit of measure.* If I come to you and say that a bottle of water is "two," you'll rightly think I'm an idiot. But if I tell you it's "two liters," you'll understand that I'm talking about its capacity, not its price or weight.

Having a unit of measurement actually means first of all having a precise initial reference (the "liter," something we know well), near which we then define our quantity. Which, agreed, is something we probably all know from elementary school. So why reiterate it?

The answer is simple, and you probably already know it: because we tend to be "a bit dumb," and thus the distortions of our mind will constantly lead us to focus:

- **On "irrelevant units of measure":** Maybe the only thing that matters in a purchase is *the cost of an item in relation to how often we will use it,* but we focus on a multitude of useless things such as its size, weight, design, etc.

- **On "non-relevant" scales.** That is, on how large or small those numbers appear next to benchmarks that, pragmatically speaking, are of no use to us! Think of the example given a few pages ago when we talked of comparing our annual salary with that of the world's richest individuals. We don't want to, yet our minds will end up focusing on the fact that our salary is a tiny fraction of Bill Gates' (tiny fraction! Insignificant amount! I'm so poor! I can't do anything about it!), rather than a +12% increase from last year (nice increase. I'm doing well! I can keep doing what I've been doing).

But let's also consider the different contexts where we are led astray by practices and common sense, as typically happens in the case of measuring our physical fitness: we often tend to associate it with possible weight gain or loss, when in fact, we should link weight variations to measurements of the ratio between fat mass and muscle mass.

The message here, in short, is once again as simple as it is necessary: if you need to evaluate numerical data, always do so *through the lens provided by the unit and the scale of values relevant to the context of your interest.* Do it always in relation to the objective you are pursuing, and hence in such a way that it serves as the best theoretical indicator of the value of different possible strategies. The rest is typically just distraction.

> *"Perhaps in your company, beyond revenue, it is necessary to evaluate the environmental impact and the satisfaction of your employees. Perhaps, following that calamity, beyond the undeniably important number of deaths, it is necessary to also assess the negative social and economic impacts that it will have on the region in the long term. Perhaps in your romantic relationship, beyond the number of texts your girlfriend sends you daily, it is important to understand if and how much she values you as a person. Perhaps in your child's report card, beyond his performance in various subjects, it is necessary to consider... the skill of his teachers! [...] An example I like to give in relation to this discussion is that of the "report card" given to a student mid-year. We often tend to see it as a measurement of his academic effort, forgetting that there is a teacher on the other side who may have developed a*

poor teaching method. And this is precisely where the revolution of " measuring what no one else would measure" could come in: an evaluation of the quality and goodness of the teacher's work by the students, for example. It would help to understand the true source of the problem and, consequently, to find a much more effective solution."

Quantity and quality

In short: where possible and convenient, *increase the quantity and quality* of your measurement tools. In other words, simply, try to improve the "sources" that provide you with the data you're interested in. Imagine having to choose which movie to watch at the cinema one evening, and wanting to determine which might, in theory, appeal to you the most. It's not exactly a life-or-death situation, but I find it interesting precisely because of its complete "unscientific nature" and the fact that it can fall into the category of topics of practical interest discussed at the beginning of the book: *everyday problems arising in contexts not dominated by hard sciences*, yet where *applying a method can still help us extract more statistically reliable rules*. This can obviously lead to a few different problems, like misinterpreting the data due to personal biases or incomplete information, ultimately leading to suboptimal decisions. But let's delve into this one step at a time. We'll explore strategies to mitigate some of these issues, while for others, *we'll accept their limitations and focus on minimizing their impact.*

And so, in a situation like the one above, you could start by reading the articles on that film review page you trust so much and plan your evening simply based on which movie has received a high numerical rating.

But considering that so far it might have happened at least once in, let's say, three times that you didn't agree at all with the tastes of the reviewer working for that site, you could expand your strategy by increasing the number of your measurement tools, namely:

- Gather five or ten movie reviews from sites you consider "reasonably reliable" in relation to your tastes and "extract" the final ratings from them.

- Now, assuming they are on the same scale (all from 1 to 10 or from 1 to 5), calculate the *arithmetic mean* of these scores. And if you don't remember what that means: add them up and then divide this sum by the number of values you have added. For instance, if you have 1, 2, and 4, it will be 1 + 2 + 4 = 7, divided by 3 values = 2.33 repeating. Or if you have 2, 4, 6, 8, it will be 2 + 4 + 6 + 8 = 20, divided by 4 = 5.

  And what if they are not on the same scale? Here too, those who already have a basic arithmetic mindset might already answer the question and move on. For the others: you could *convert them all to the highest scale*. That is, take the rating with the higher maximum (in the example above, it would be the one from one to ten compared to the one from one to five), divide that "higher maximum" by the maximum in the rating you are considering (which in the one to five measurement would mean 10 / 5 = 2) and multiply the result just obtained by the rating relative to that maximum (if the considered rating, for example, was three stars out of five, this would equate to 2 x 3 = 6 out of 10).

In short, this mean will be the "final grade" for your film, and its partial reliability lies in the fact that this arithmetic tool tends to *reduce possible statistical fluctuations* in individual measurements. For instance, if one reviewer happens to have a bad night and ends up giving "one star out of ten" to the masterpiece of the millennium just because they were in a terrible mood, the increased size of the statistical sample will effectively end up reducing the "weight" of that random event. It's for this same reason that, as we'll see later, whenever the result of any laboratory experiment might be influenced by unavoidable random factors, a good solution to adopt is to repeat the experiment as many times as possible.

Obviously, this leads to another equally important truth: an experiment must be conducted with "reasonably rigorous" criteria and tools. It wouldn't make sense, for example, to weigh oneself in the morning if the scale we use is completely broken, and it's probably not advisable to try to fit a shelf into a space measured "by eye." In short, during the measurement phase, we must never forget the "qualitative" factor of the instrument we use, meaning the reliability (at least potential) of the number it "produces" in relation to the reality we are interested in.

How many times, for instance, has a critically acclaimed film put you to sleep in your seat? Clearly, in "non-scientific" contexts like choosing a film, *achieving absolute reliability is impossible.* And looking for this reliability where there isn't often leads to analytical disasters and pseudoscience. However, let's consider once again that we are simply trying to make the best decisions using statistically predictive criteria and let's continue.

So, in order to "increase source quality" in such a context, we could embrace the concept of "trust" in specific reviewers. By focusing on the opinions of four or five critics whose preferences consistently align with ours, we might achieve a more accurate predictive value than by averaging twenty-five randomly chosen ratings. This idea also ties closely to the concept of *weighted average,* where not all inputs are treated equally. By assigning greater weight to sources that have demonstrated reliability or alignment with your tastes, the resulting prediction can become significantly more meaningful. It's a method worth considering—and perhaps worth experimenting with yourself.

A last thing here, before moving on: this idea of "increasing sources to reduce statistical fluctuations" immediately reminds us of the importance of *community* in evaluating the outcome of scientific research. It is in fact often heard of a doctor with an "outsider" opinion who reveals, in contrast to the entire scientific community, that in reality the disease "x" can be cured with the miraculous remedy that no one talks about; or, conversely, that a drug with well-established efficacy is suddenly extremely harmful.

This is where the community plays a role: by comparing the outsider's data with that from other labs and centers to ensure accuracy and reliability. Fundamental claims, such as those concerning human health, are never left to arise from statistical fluctuation (or, as often happens, from bad faith) of the individual "study-measurement tool" but go through numerous cross-checks, until this data is confirmed as relevant if it continues to appear, or dismissed as a "phantom" data, as a result of an error from the individual source. It is clearly a complex process, in which answers can often remain incomplete or unsatisfactory; but what matters, as usual, is that given a sufficiently well-designed model, it will likely lead to the best "model of reality" conceivable in that context. Then it will be necessary to accept: 1) That older models may (and in most times *will*) need occasional updates, and 2) That more complex facts will require time and effort to emerge.

Estimations

All the discussions we've had about movie nights should have already hinted at what we'll talk about here: where you cannot or it isn't practical to measure precisely the quantities you're interested in, it can still be very useful to tighten your "quantitative grip" on the situation at hand through *estimations*.

In other words, when this is more convenient on an effort/reward basis, trying to replace measurement (which maybe isn't possible or worth it at the moment) with a quantitative estimate could still be more useful than completely abandoning numbers. For example, if you are trying to follow a weight-loss diet, even if you don't calculate the exact calories you're consuming per meal, it could still be acceptable in certain cases to determine whether you're ingesting about 400-500 or about 800-1000 calories per meal. Or, to approximate even further, whether you're eating more or less than the previous day (a different unit of measure, but still relevant).

Clearly, as we have already seen a few pages back, inaccurate measurements can lead to real challenges. The less a numerical quantity is extracted with precision and rigor, the more approximate the information about reality will be. Consequently, the more you need to make decisions in a field that is risky and governed by predictable laws, the more you must abandon your estimates and instead strive to measure as precisely as possible. Consider the engineering principles behind the design of a bridge or an airplane: the positioning of every single component must adhere to precise physical laws, and the risk of ignoring them would be far from negligible. Thus, the precision of every quantitative assessment becomes crucial in making the difference between success and disaster.

All these considerations on "risk" and "scientific rigor" should by now be more or less clear, but I keep reiterating them for one simple reason: because during our days, *we constantly forget about them*. If there really was no longer any need to stress the importance of "not ignoring numerical realities where they leave no room for doubt," then why do families, companies, and even entire countries continuously end up going bankrupt? And if it's obvious that continuously consuming more calories than we burn results in body fat accumulation, why do some countries like the United States still have an alarmingly high obesity rate that poses a significant concern for public health? The privilege of formulating an answer is yours.

Strategy Lab - The Mad Numerator

This strategy lab can help you move evaluations from the usual "fuzzy vagueness" of the mind to the "correct rigor" of quantitative assessment. The following list of questions can be used both as a pure brainstorming exercise and as a useful checklist for any project in which you are trying to design your "scientific strategy."

Remember: asking yourself questions isn't about finding immediate answers. It's about embracing a "scientific" mindset where a question serves as the starting point for exploration. The interaction between your mind and the question—the tension between the unknown and your curiosity—creates the spark for your next steps. Accept the effort that any intellectual process requires, and trust that meaningful truths take time to emerge. Understand, too, that initial answers, much like in science, often lead to deeper questions. And if some the list feels heavy as you read, don't stress: skim the questions for now and revisit when you're ready to apply their investigative approach:

- How important is the goal I've set for myself on a scale from 1 to 10? And how will it truly impact my life in the long term? Is there a discrepancy between these two quantities?
- Am I avoiding measuring something important? Should I measure more, or perhaps something specific?
- Am I considering the wrong unit of measurement? The wrong scale?
- Am I ignoring quantities that could actually prove to be essential?
- Can my strategy simply be based on raising a specific number? On reducing it? On making it irrelevant in the given context?
- Can my strategy simply involve intervening to alter the rate at which a quantity changes.
- Do some problems arise from having measured, or assumed, a completely wrong quantity?
- Are there elements whose "weight" has been measured and quantified incorrectly due to a bias in the procedure? If so, why, and how could I fix this?
- Can I measure something that no one else would measure?
- Did I make estimates that were too rough to make good decisions?
- Have I considered how the measurements will vary over time? Will they do so reliably? Predictably? How might possible variations "impact" my project?
- What happens if the variables change in the "worst possible way"? Is it a real possibility? Am I prepared?

'I believe there are 15,747,724,136,275,002,577,605,653,961,181,555,468,044,717,914, 527,116,709,366,231,425,076,185,631,031,296 protons in the universe,

and the same number of electrons."
(Sir Arthur Eddington)

Stories of "Scientific Champions"

Richard Feynman's Quantum Dream

The world of physics might appear to most as an intricate maze of equations. Yet in the mid-20th century, there was a man who navigated this maze with unimaginable ease and enthusiasm. His name was Richard Feynman, and his passion for research and science transformed not only his field of study but also the very way we see the world.

Feynman was born in New York in 1918, into a family of Jewish origin. As a boy, he was already fascinated by mathematics and science. After completing his Bachelor's degree at MIT, he moved to Princeton University, where he began studying subatomic particles.

But it was in the post-war period that his true passion, quantum physics, found its outlet. Quantum physics (for those who may not know) is the study of phenomena occurring on incredibly small scales, at the level of atoms and subatomic particles. It was a field full of mysteries, but Feynman was not one to be easily discouraged.

With his unique approach, combining intuition, mathematics, and a deep curiosity, he developed what is now known as the "Feynman diagram formalism," a new way of visualizing quantum interactions. These diagrams literally revolutionized the field, making complex concepts accessible and visualizable.

Feynman was also known for his skill in teaching. His lectures were legendary, blending personal anecdotes with profound scientific truths. He firmly believed that knowledge should be shared, and his unique style inspired generations of students and researchers.

His impact on the world of physics was, quite literally, "irreversible." For his contributions to the theory of quantum electrodynamics, he shared the Nobel Prize in Physics in 1965.

Moreover, behind the scientific genius, there was also a man with deep emotional sensitivity. After the death of his beloved wife Arline, Feynman continued to write letters to her for years, putting his thoughts, hopes, and fears on paper. These letters, filled with love and longing, reveal a vulnerable side of Feynman, a side that perhaps only a few knew.

The story of Richard Feynman exemplifies that science, far from being merely "cold" and rigorous, often symbolizes passion, enthusiasm, and curiosity—fundamental elements that inspire and illuminate the world. At the same time, it shows us that even the brightest minds may need their rituals, even if they seem irrational. Sometimes, great science and the intimate human desire for ritual and storytelling can coexist, and perhaps it is this very interaction that gives rise to the most interesting and fascinating reflections on human nature.

II - Mathematical Distributions and Buddha

Summarizing what was discussed in the previous chapter: a "reasonably" thorough and objective understanding of reality can be a crucial strategic tool for aspiring, at least in theory, to success, effective action, personal well-being, and balance. Once we "extract" reliable information from this reality about its laws, characteristics, structures, and behaviors, we can also try to understand how to best intervene in it.

This has also led us to several problems. One of the most important among them: when faced with the complexity of certain practical situations, the ability to "know scientifically" and thus develop effective solutions can become quite a challenge; alternatively, in some cases, it might simply prove impractical or pointless. Furthermore, we must inevitably add: even perfect knowledge of a given context might still "not be enough" to overturn a particularly unbalanced power dynamic.

And since we will talk more about the second issue later, let's focus for a moment on the fact that often, to investigate a phenomenon, one must necessarily narrow down, simplify, isolate, and therefore *reduce the complexity of the problem to a manageable level.*

This leads us to another fundamental definition within the "scientific strategic theory" we want to discuss, namely the concept of an *aspect* of a situation, which we might express as: *"any of its components, or a set of components, grouped according to any criterion."* This is, indeed, a somewhat pompous way of saying that *anything* can be an aspect of something else, as long as there is at least a partial relationship of inclusion, correlation, or belonging. For example, taking the situation "your kitchen," an aspect could be your table, your set of forks, the time you start cooking, and the soft light filtering in through the window on a rainy morning.

And this is a definition that, despite its vagueness, is fundamental for two simple reasons:

- We will always have limitations, both in terms of vision and resources; therefore, whatever the problem to solve, we must always *choose which specific aspects to focus on* and then deliberately "ignore" and cut out the rest.

- Despite the complexity and changing nature of certain situations, it is undeniable that their "state," or at least the state of what interests us within them, *does not depend on the totality of the aspects that compose or participate in them.* Instead, there is almost always a minority of more significant aspects that somehow "generate" the rest of the interesting effects. A person with a quick, intelligent, and trained mind is certainly not so because they actively control the functioning of all the cells in their brain, but because they probably maximize their cognitive performance through the right practices of health and prevention. A country is at war not because every man, woman, child, and animal shares the war cause, but because this cause is perpetuated by its highest spheres and the army they command. You wake up in the morning and feel good about yourself not when every single thing in the world out there is going well, but when those two or three fundamental aspects of your life seem to be going just right.

It is likely that the principle defined in these last two points is known to many of you as the Pareto principle, or the "80/20 rule," a mathematical principle which, although it doesn't have a "hard" scientific basis (it is not, in short, a universal law to reliably describe every natural phenomenon), seems to have a strong probabilistic/statistical foundation.

Joseph M. Juran, a pioneer in the field of business management, was the first to recognize and apply the observations of Italian mathematician Vilfredo Pareto, transforming them into the so-called "80/20 Principle." In his initial 1896 publication, *"Cours d'économie politique",* Pareto had highlighted this analytical phenomenon by noting that 80% of Italian land was owned by 20% of wealthy families. Since then, many studies, such as those by M. E. J. Newman, a professor at the Department of Physics and Center for the Study of Complex Systems at the University of Michigan, have confirmed that numerous natural phenomena and complex systems adhere to this "power distribution" law—an empirical law where a few significant elements tend to control, modify, and redefine everything else. For instance, a famous Microsoft report observed that fixing the most frequently reported 20% of bugs could resolve 80% of system crashes and errors. Similarly, in the field of health and safety, it has been determined that roughly 20% of risk factors cause 80% of issues.

From all this, we can derive the extraordinarily useful rule of thumb that every concrete goal can be achieved by focusing solely on what we will now call "key aspects" because:

1. More *important, significant, impactful* in the given context

In short, that "20%" (which could be a 10%, 15%, 30%...) that is able to maneuver, control, and manage the other aspects of the situation that interests us, doing so by virtue of the goal we have set for ourselves.

2. More *immediately or easily manageable* or controllable

There's no need to break the lock on your front door when you've forgotten your keys, if you can simply turn a handle and enter from the back. There's no need to alter a video game's programming to win it, when you can just pick up the controller, play it, and learn the right strategies to defeat its bosses. And these are all "incredibly trivial" examples of the principle, yet the concept of "seeking the most controllable or changeable elements" remains, in my opinion, something incredibly underestimated in daily life. Just think of all the times we insist on taking a difficult and convoluted path when there are easier and more direct alternatives. Or when we cling to an idea, a job, a habit, or a relationship that no longer benefits us, instead of seeking new opportunities and open spaces.

3. Numerically *essential*

So chosen in such a way as to be sufficient but not excessive. There's no need to intervene in twelve areas when you can achieve the same result by addressing just two. Obvious? Yes, but this, aided by our "biases," *is perhaps one of the most overlooked truths in the world.*

If you want, you could even finish reading the book here, because from what has been said so far, it can essentially be concluded that every conceivable "successful strategy" is based on this principle: *identify and "isolate" what objectively seems to matter most; and then, apply your resources and energies there.*

However, if you should decide to continue (which, of course, I sincerely hope), let's make some additional considerations on the matter. First of all, we might observe that a big part of the beauty of this approach lies in inviting us to adopt that typically scientific mindset of "calm down and rationalize": even when urgencies appear, when stress and pressure rise, or when the problem seems enormous, trying to focus solely on the *smallest possible number* of essential key aspects is also a way to avoid being overwhelmed.

Calm down. Analyze. Gradually narrow down the context to work on. Isolate and concentrate on the very "fundamental bits," set yourself one micro-objective at a time, and you might find that the immense space station capable of destroying planets and terrifying the galaxy can be blown up with just a well-aimed shot down an air duct.

Strategy Lab - 80/20 Questions

Given what has been said so far about the Pareto distributions of power laws in complex systems, let's try to provide a list of questions aimed primarily at identifying the "20% that controls the remaining 80%" in the situation that concerns us. This does not necessarily mean aiming to collect all the required answers immediately, but rather to find in these questions a set of guidelines that helps understand where it is beneficial to investigate.

In this case, for example, we might ask ourselves:

- Which element here is most important, as it is the one most capable of generating significant effects for me within this situation?
- Which elements, if changed, would completely alter the situation?
- Which seemingly irrelevant elements might play an important role?
- Which element holds the greatest power? Which one can bring the most benefit?
- Which element can cause the most damage?
- Which aspects can be easily eliminated, modified, or controlled in order to turn the situation to my advantage, perhaps with the maximum result and minimum effort possible? For example, who has already addressed such aspects? How? What did they achieve?

- What generates this thing or situation? What controls it?
- Are there "states," particular evolutions of the situation where I could maximize benefits or minimize damages?
- How can I take advantage of this when it or part of it becomes removable, controllable, or manipulable according to what I desire? How can I make it become what I want it to be?
- How can I transform part of it into what I desire?
- Should I remove or weaken what's negative? Empower what's positive? Add new positive forces?
- Is it possible that I'm focusing on a set of less critical elements, when just a fraction of that effort on more important ones would suffice?

Strategy Lab - Three Approaches

3 Key Aspects: Take the situation you want to address and, perhaps starting with the questions from the previous "strategy lab," identify three key aspects to focus on. No more, no less. Ensure that the elements considered adhere to the principles above (essential/important/manageable), and commit yourself to work solely on those.

An alternative exercise you could do here is: choose a situation you're already dealing with daily. Perhaps a current project, a job you're dedicated to, or a problem you're trying to solve once and for all. Examine the "aspects" you've decided to concentrate on to complete your path: *are they essential, or are you engaging in something redundant? Are they important, or are you focusing only on something "attractive"?* Are you working on something manageable, or wasting time, resources, and energy hitting a "wall" that will not produce any useful results?

Tug-of-War. The most interesting thing about this technique is its simple and intuitive way of modeling any situation (it's no coincidence that similar principles have led to various analysis tools in business management, such as the SWOT matrix): two factions pull a sturdy rope in opposite directions, one in favor of your objective, and the other acting as an antagonist. And finally, there's you, doing everything possible to reduce the strength and number of people pulling from the opposite side, while increasing the pull of the side that interests you. This is not too far from reality if you consider that the pulling forces may not be muscular but are still sets of forces of a chemical, physical, thermodynamic nature, and so on. Therefore, list two negative aspects and two positive ones of the situation you're interested in. Then, try to work directly on "reducing," "avoiding," making the former ineffective and "supporting," "embracing," the latter. Now try again, and do it at the "highest difficulty level."

Also, if you had to choose ONE positive aspect and ONE negative aspect to act upon, what would you choose?

Make what is manageable important, or make what is important manageable? This approach can serve as an alternative or complement to the "tug of war" method. Simply, as suggested by the title: once you have identified the "key aspects" of the context that interests you, instead of basing your goals on "boosting" the negative and "weakening" the positive, aim to enhance the power, vigor, weight, and importance of what is more directly under your control, and to increase the control, manipulability, and access to what already holds power, weight, or importance. Through this approach, you might come up with a bunch of potentially effective and interesting strategic lines.

Everything has a limit

As we have just seen, collecting information can begin by appropriately "defining" the portion of reality that interests us.

This allows us to understand in detail where, specifically, it is most beneficial to apply our efforts, resources, or attention.

Yet, even in applying this first, simple step, one can encounter initial problems: when defining a context, you might consider a portion of reality that is either *too limited or too broad*. This leads to obvious consequences: in the first case, you risk ignoring useful elements, and in the other, you might disperse your energy and resources analyzing fields that are too vast. For example, imagine you are working on a software project, aiming to develop a revolutionary social network app capable of dethroning all industry leaders, from Facebook to Instagram, including YouTube and TikTok. What can a "more scientific approach" suggest in this scenario? Should you first examine the market? The competitors? The latest technologies not yet fully exploited in the field?

Clearly, the answer here is "fairly straightforward": much depends on the time and resources available (typically more resources and more time = more exploration), the risk factor present in the field (more unsustainable risks = more consideration of conventional or known paths), as well as the approach method one wishes to employ (more creative? More pragmatic? More "economical"?). The following exercise, for example, can serve as an excellent guide in this regard.

Strategy Lab - "In and Out of the Box"

Once you've established your goal or the problem to solve, aim to find two solutions:

- **"In the box,"** or within what is known, obvious, and conventional. This, related to what was mentioned earlier, simply means "defining the scenario" to what is "common knowledge" or previously known within that field. For example, are you dissatisfied with your job? Develop a solution simply based on going to your boss and discussing what you find unsatisfactory, and leaving if the negotiation doesn't lead to a positive outcome.

- **An "out of the box" approach**, which obviously means doing the opposite: if you usually "narrow down the scenario" to a point, try instead to focus on a "horizon" that is completely unknown, original, and surprising to you. For example, if again, you are not satisfied with your job, try to figure out what quirky interests you could pursue after work to best relieve tension; or consider the idea of doing exactly the same job but remotely from some tropical island. When you think "out of the box," simply don't limit your imagination!

Then, for each developed solution, try to quantify (or estimate) the potential risk and the cost in terms of relevant resources (Money? Time? Cognitive energy?). Choose whatever scale of values you prefer, as long as it is the same for every potential plan considered. The purpose of adopting this technique is to learn how to effectively balance these parameters, and even to mitigate the influence of any bias that would divert us toward either too narrow or too broad fields. For example, the desire to pursue an original thought "at all costs" might lead us to overlook the fact that sometimes the most obvious and common solutions are the most sustainable ones. Alternatively, forcing ourselves to think "outside the box" can help redirect our attention precisely to those unconventional solutions that we would normally ignore out of fear, insecurity, and a desire to avoid conflicts or changes.

The Secret of the Super Lazy Ones

Alright, so... study, research, analysis, "measurements." Identify the key aspects and focus solely on those.

But... what happens if we need to understand a situation more deeply and we don't have time to analyze it, study it in detail, and comprehend all its aspects and dynamics? After all, this happens every day: we certainly can't take a macroeconomics course every time we want to invest our savings, nor earn a medical degree every time we have a headache. As Socrates maintained, *wisdom comes from an awareness of our ignorance;* yet at the same time, the time available in a day is barely enough to do the essential things to "get by." So how do we think of "bridging" this profound ignorance and still take effective action outside our field of expertise? And the answer, I'm sure some of you have already guessed, *lies in relying on intelligent listening, or intelligent imitation, of those with more experience.* This concept has been with us since childhood. Think about the traditional learning model, as seen in school or university: the "learning by listening" approach is represented by teachers, mentors, lectures, and seminars, while the "learning by imitation" approach emerges in internships, apprenticeships, and simulations. Similarly, in the workplace, we encounter these methods through superiors, experienced colleagues, and hands-on workshops.

In essence, we could differentiate between two fundamental approaches (which, of course, are often entwined) to gain a deeper understanding of the world: one is *analytical,* based on examining the aspects of reality and theoretically more resource-intensive, albeit with higher potential. The second approach, on the other hand, relies on using *"pre-packaged knowledge,"* with all the obvious limitations that come with it.

> *"According to a study conducted in Great Britain, during which a competitive multiplayer 'survival game' was simulated, and published in the prestigious scientific journal 'Science,' the strategy of 'imitating the best' proved to be one of the most effective strategies overall for competing in an environment, regardless of the level of hostility and complexity of the context in which it was applied."*
> **(From the book "The Art of Winning Unfairly")**

Let's pause for a moment to consider what was just said about approach number two. First, let's focus on the concept of "listening or imitation" as, in turn, this distinction leads us to classify two main categories of cases: one where the reference figure becomes *a source of information* (such as when a doctor needs to suggest a certain therapy), and the other where the same figure becomes *a role model* (such as when a small entrepreneur wants to grow and starts to imitate the strategies of Steve Jobs and Jeff Bezos). It's clear that these cases can overlap and intersect in practice; however, the important thing is to understand their common characteristics as an alternative to "recreating" certain knowledge from scratch.

First objection someone might raise here: *"But experts can be wrong!"* Well, obviously. However, the response to this, in my opinion, cannot be the "total denigration" of the expert figure or the belief that anyone with advanced knowledge or experience is some sort of "guru" residing in their ivory tower. In recent years, global political rhetoric has increasingly relied on the "denigration of graduates and professors" as fuel for easy populism. While this approach verges on the criminal for fostering a kind of "intellectual dark age," its success cannot be denied. This success likely stems from its appeal to those who have abandoned intellectual pursuits and, in doing so, feel entirely "absolved" of personal responsibility.

To truly restore the value of study, effort, and work, we must dismantle this rhetoric and reestablish respect for the role of the expert. While experts, like anyone, can be fallible or subject to bias, their deep understanding of a field's complexities makes them the most reliable source of informed and potentially valuable insights. Of course, as mentioned, we could choose to start from the very basics, study the principles ourselves, and analyze the field of impact in depth. However, this approach often demands a level of time, effort, and resources that far outweighs the potential reward, making it an impractical path for most.

That said, learning effectively from experts or role models requires more than passive acceptance. It calls for what we might call *"smart" listening and imitation* and involves a series of important considerations:

Information from the expert should be "weighed"

We mentioned this a little while ago and will reiterate: the expert can make mistakes. More importantly, it is easy to observe that even in the most rigorous fields, *there are experts who say things that are literally in contradiction with each other*. We discussed this a few pages ago, facing the dilemma of "if multiple sources say different things, where is reality?" In truth, this is a philosophical problem that is not easily resolved and stems from the fact that some realities are just extremely complex. However, we might try to adhere to guidelines such as:

No "ad personam." Not long ago, we saw the first rule for improving the quality of our perspective: becoming aware of our own biases and avoiding falling into them. So it's very easy for us not to listen to the arguments, even if impeccable, of someone we particularly dislike, reminds us of our alcoholic uncle, or holds a different political view from ours. Conversely, we will instinctively tend to warmly embrace the nonsense of someone we deem "good and capable." Therefore, even in this case, as good scientist-strategists, we must make an effort to notice when we apply such prejudices and evaluate solely the source that "objectively" seems to be of higher quality. This leads us directly to the next point.

Give preference to the expert *(or source)* **that adheres to all the "common sense" principles outlined in the book so far.** Love for reality. Transparency on methodology. Attention to the elements that carry the most "weight" in a situation, without succumbing to distractions and traps that draw one towards what is simpler and more attractive. The ability to detach from personal biases, such as those generated by one's own ideology or belief. A willingness to rely on numerical quantities, verifiable sources, and precise data, where the rigor of the field demands it. If an expert follows these criteria in studying and communicating their truths, and most importantly, if they are ready to accept and reconsider when confronted with realities that contradict them, they are likely more reliable than others. This cannot, of course, be an indisputable proof that this source will always and invariably be truthful, but in my opinion, it nevertheless represents a "reasonable indicator of quality."

Try to consider their risk and benefit level. In other words, an expert source may be more reliable when it takes a risk in communicating certain information, and it *may* be less reliable when it stands to gain from it.

Let's pause here for a moment, otherwise we risk making a mess. First element: risk. Take the case of a financial advisor who is providing consultancy on your behalf. If they are ready to tell you that a certain set of investments will yield a lot, that's all good. But if you manage to understand what they have in their own portfolio, you might gain something even more reliable. It's an insight for which, as they say, they have "skin in the game," which could enhance its credibility.

Let's now move on to the "benefit" part, which can be summed up by the phrase *"cui prodest"* from Marcus Tullius Cicero's time (1st century BC; although the expression, rendered as *"cui bono,"* is first attributed to Lucius Cassius Longinus Ravilla, 2nd century BC). In other words, the question to ask is: *who benefits when the expert provides us with certain information?* Because, yes, there are indeed scientists who, acting in bad faith, disseminate research results significantly skewed by funding from specific industrial groups (a phenomenon powerfully illustrated in the TV series Dopesick). However, the first step is to approach such situations with discernment: evaluate each case on its own merits, identifying those instances where conflicts of interest or the pursuit of personal gain are blatantly evident. When then such influence is clear and uncovered, it is critical to challenge the integrity of the particular group of scientists or that specific research, *rather than dismissing the credibility of all experts*; or, worse, undermining trust *in the scientific method itself.*

The art of questioning through *cui bono* should not become an excuse to turn us into conspiracists who believe that behind every attempt to gain a benefit there lies a malevolent architect ready to shamelessly lie. Not to mention that, yes, one might even encounter an expert who, despite benefiting, is still telling the truth. Consider the case, certainly paradoxical but effective, of a scientist paid by the fruit and vegetable lobby. Even if he is "corrupted" in shouting, "Fruits and vegetables are good for you!" fruits and vegetables will still be good for you.

In short, as usual, it's permissible to start with this principle to raise a doubt. However, reality is complex, and therefore accepting only the facts that confirm our initial doubt while continuing the investigation would be completely unscientific.

Compare and contextualize. Given the criteria outlined so far, you may find yourself deciding "beyond any doubt" to discard some types of information sources. However, there will be many other times when you are faced with multiple sources that, although conflicting, each present something valuable. For example, if one doctor claims that their treatment is the most effective for a certain illness and presents their data, and another doctor presents similar data with a different medication, it's possible that *both could be right in context;* perhaps because there are additional factors, previously unconsidered, that allow both truths to be "unified" within a broader, singular reality. Whether we can carry out this process on our own is uncertain, especially if we are dealing with fields that fall outside our expertise or knowledge. The "golden principle" to be extracted from this discussion is: keep in mind that sometimes seemingly contradictory realities can coexist and both be "true," and this applies to both common everyday principles of good sense and the most rigorous data from scientific research. Therefore, an exploration in this sense, a search for this "higher truth," might require us to undertake the immense challenge of *carefully balancing and comparing different sources.* This might be completely beyond our capabilities, but when it comes to academic research in any field, it's likely that a subsequent study or a meta-study will step in to perform this process for us; and yes, for those who are unaware, meta-studies are precisely "research of researches," comparisons and contextualizations of data aimed at "bringing forth greater truths" based on individual studies.

Therefore, a true "ultimate trick" to obtain more reliable knowledge on any scientific issue is: seek out the most recently published meta-study in a reputable journal that deals with that specific topic. Assuming we have the necessary foundational knowledge to read and interpret such meta-studies, the "magic" often lies there. Additionally, with the advancement of AI and LLMs, it may become possible to perform this highly complex data comparison with much greater ease, in a way that is accessible and even explainable in a form that everyone can understand; or perhaps this is already possible at the moment you are reading this.

> *"When an elderly, distinguished scientist says something is possible, he is almost certainly right. But when he says something is impossible, he is almost certainly wrong."*
> **(Arthur C. Clarke)**

You can always try to prove the expert wrong, but with tools of equal or greater rigor than theirs

You have probably already noticed that out there is far too full of people trying to debate medicine or macroeconomics, not only without having a degree, but even without ever having delved into a book or any other informational medium that could illustrate the complexity of the topic or the rigor needed to piece together certain truths. You have likely heard of the "Dunning-Kruger Effect," the mental bias that leads individuals lacking particular knowledge or skills to overestimate their abilities. Much of this can also be attributed to our human need to design *simple answers*, which allow us to reduce cognitive stress.

After all, the daily harm caused by Dunning-Kruger-affected individuals—both in public debate and to themselves, such as when people self-diagnose or use ineffective treatments for serious illnesses—underscores a key principle: in fields governed by "hard" or "semi-hard" sciences, where complexity is high and rigor is crucial, *challenging expert assertions is valid but must be done with the same rigorous methods and tools used to establish those truths.* Laboratory vs. laboratory. Complexity vs. complexity. Rigor vs. rigor. Otherwise, it's a gamble: acceptable only if taken consciously and weighed against the potential cost of being wrong.

This also highlights that knowledge and experience in fields not strictly scientific—such as art, certain social sciences, and similar areas—stem from intersubjective, relative, or non-scientific foundations, making them further removed from the "objective reality" we often reference. As a result, they can be much more easily "disassembled" and replaced by "outsider" perspectives. In these domains, even a newcomer with minimal prior knowledge can sometimes grasp or create ideas that are unexpectedly functional, valid, and impactful. The further one moves from the rigor of data, the more "anything and its opposite" can hold true. Here, *"it must be this way because that's how reality is"* gradually gives way to *"as long as it works."*

For instance, if an experienced writer insists a book must follow specific conventions, but you break all their rules and sell countless copies, why not? Or if an Olympic gold medalist outlines their precise training regimen, yet you develop a completely different approach—perhaps based on deeper insights into the human body's nuances—and break their record, why not trust your method? So, *pragmatism* comes again into play in how we implement these ideas, proving itself as *one of the most effective ways to extract "signal" from the "noise"*; and, consequently, as an incredibly powerful tool for crafting a "rational rebellion" against those experts who actually fail to see beyond their own narrow perspectives.

Strategy Lab - "Follow the Buddha and Then Kill Him"

Let's specifically return to the sub-case of experts used as "role models" rather than a "source," and consider a famous Zen saying that roughly states: "If you meet the Buddha, kill him." This is obviously not an invitation to gratuitous violence against random role models, but rather a precept that should encourage evolving and emancipating oneself from one's teachers. Starting hence from the lesson offered by this sentence, let's try to put together a strategic technique that applies part of what we've seen so far, integrating it with a "dash" of new tools.

Find your Buddha. Quite simply: when you want to reach a certain goal, likely someone has done it before you; or, even if they haven't fully succeeded, they might have developed significant expertise in the process. Therefore, research, investigate, examine, and try to select your ideal "master".

Then analyze this information with the right "scientific eye" (cleansed as much as possible of "bias") and try to extract cause-effect relationships, key strengths, and strategies from the story of your "Buddha." Additionally, in the event of contradictions, doubts, or uncertain sources, always remember to filter this information based on what was mentioned a few pages ago: favor the "inconvenient" ones, those issued (at least apparently) in good faith and transparency, and those whose authenticity the "Buddha" would probably bet their own skin on.

Don't get discouraged. Be rational. Let's assume you have enough information now. However, information alone might not be enough, and it's still possible that your goal is truly ambitious, and the "level" reached by your Buddha might be very difficult to achieve. Obviously, once again, the best advice is the one suggested a few pages ago: "step off" the emotional evaluation train, try to think like a "scientist," and start thinking more quantitatively. To achieve a goal worth 100, someone might have used two segments of 50 that you don't have. But if instead of getting discouraged you start checking your pockets, you might realize you have enough "pieces of 2" to at least reach 90. Consider it as if it were forces converging at a single point. Even if you cannot precisely replicate the results of a specific path, there must be some actionable insight, a guiding rule that can empower you. Most importantly, there's always a way to compensate for any initial "disadvantage" you might have in comparison to others. Can you offset lesser genius with greater creativity? Fewer initial resources with greater daily effort?

The truth is, you cannot know for certain if you'll reach your "Buddha." And more often than not, your own biases will convince you that you won't. But for this very reason, let go of the worry and focus entirely on how to leverage what you do have to get as close as possible. The rest is nothing but an over-spiced "stew" of biases, likely to lead to cognitive "indigestion" you're better off avoiding.

Start thinking "as if." Even if the idea doesn't fully convince you, give it a try: start acting immediately as if you were your "Buddha." Try to step into the "mental state" you imagine he might have been in when facing his own battles, and let that guide your actions. How do you picture him? Resolute? Determined? Calm? Fierce? And how do you envision him responding to the challenges you face?

Let's be clear: merely thinking "as if" isn't enough to develop real qualities. This "as if" mindset should not in fact be your final destination but a starting point—a tool to project yourself, even if only ideally, into the experience of "being your model." It's a way to get closer, to better understand the challenges, obstacles, and problems that come with embodying that ideal.

Above all, it's an opportunity to embrace a bit of naive recklessness —some bold, foolish experimentation that might yield valuable lessons. These experiments may not always succeed, but they can bring you closer, faster, to your ideal endpoint. And above all, let this playful audacity become part of your journey.

Kill your Buddha. Don't just try to imitate your model. Try to surpass it. Yes, even if it seems strange or impossible, give it a try anyway! For instance, try asking yourself:

- What is the "missing element" my Buddha might have overlooked?
- What are my Buddha's weaknesses, and how could I address or adapt to them to gain a strategic edge?

- What actions or approaches was my Buddha unwilling or unable to take, and how could I outperform him by working harder or smarter?
- Where did my Buddha stop or hesitate, and how can I push one step further?
- What could I notice or acknowledge that my Buddha ignored, dismissed, or failed to see?
- Which underestimated, discarded, or forgotten elements could I turn into valuable resources?
- How can I leverage unique qualities or advantages that only I possess?
- In what ways can I differentiate myself from my Buddha to carve a distinct path?
- What could I create that surpasses my Buddha's work in originality, functionality, or appeal?
- What limitations will always exist in comparison to my Buddha, and how can I compensate for them with alternative strategies or resources?

Strategy Lab - Swarm

As mentioned at the beginning of the chapter, not all systems are subject to power distribution laws of the Paretian type. Consider a swarm of ants working to "attack" a piece of bread that we have accidentally dropped on the ground: a perfectly distributed, decentralized system, coordinated (to simplify) by a hive-mind, in approaching which it is not possible to "think in terms of 80/20." We cannot think, for instance, of finding a leader ant to communicate with (assuming we knew how) to dissuade the entire group from their current action. Similarly, there likely isn't any other "focal point" on which to concentrate efforts.

If, therefore, it is true that it can be useful and advantageous to approach most problems by assuming there is an 80/20 power distribution and thus focusing on the so-called "key aspects," it is equally true that it is necessary to understand, at least in broad terms, what to do when this possibility is not considered, or it is simply unclear whether it is or not. This could be summarized in the following guidelines:

Decentralize your approach: We have observed that when faced with a swarm-type problem, one cannot hope to find a single leverage point to resolve the situation. Therefore, it is essential to adopt a decentralized approach, similar to how the swarm itself operates. This might mean creating various action plans to tackle the different aspects of the problem.

Maximize effectiveness: Although the decentralized approach requires more effort and resources, it is possible to increase its effectiveness by focusing on the swarm's most critical components. This means trying to identify those key areas where, if addressed, could have a larger impact on the system as a whole. In other words, as outlined in previous pages, even if we cannot identify 80/20 aspects, there might still be 55/45 ones to consider.

Iterate and adapt: Due to the potentially changing and dynamic nature of swarm-like scenarios, it may be more crucial than in other cases to implement a feedback and adaptation process. This involves constantly reviewing action plans, assessing their impact, adapting them to new circumstances, and continuously striving to identify focal points where most of one's energy and resources can be concentrated.

Stories of "Scientific Champions"

Heisenberg and the "Rational Overcoming of the Master"

Werner Heisenberg, a young German physicist, was in the heart of Copenhagen in 1924, eager to learn from Niels Bohr, the great master of quantum physics. Bohr was a living legend, the architect of an atomic model that had transformed the very understanding of reality. Heisenberg, on the other hand, was a young assistant, brilliant but still searching for his own path.

Heisenberg idolized Bohr. He listened attentively to every word from his mentor, absorbing his ideas like a sponge. Yet, deep down, he felt something was amiss. Bohr's theories were fascinating, but still overly tied to the classical image of the atom as theorized by Rutherford. Heisenberg, on the other hand, had already begun to sense that the quantum world was radically different, and that a further leap was necessary to propose an even more comprehensive atomic model.

One day, while walking through the halls of the institute, Heisenberg made a bold choice. He decided to follow his mentor's implicit advice—to seek scientific truth wherever it may be, even if it meant straying from the path set by his mentor. He decided to develop his own theory, guided by his intuition and beliefs.

This moment of decision was the prelude to the creation of Heisenberg's matrix mechanics of quantum theory, a completely new approach that represented particles not as small points in space, but through number matrices capable of accurately describing their behavior. It was a vision of the world somewhat different from Bohr's, and it's said that when Heisenberg presented his ideas, Bohr was initially skeptical. However, what we know for sure is that Bohr recognized the genius of the new theory, began to see it as a natural extension of his own, and thus encouraged Heisenberg to continue.

Heisenberg did it! His matrix quantum mechanics and subsequent innovations in theoretical physics earned him the Nobel Prize in 1932. More importantly, his theory became one of the foundations of modern physics, providing the essential contribution to technological developments that have literally transformed the world, from the birth of quantum computers to the new frontiers of nanotechnology.

This story is a perfect example, applied to the history of science, of "following the Buddha and then killing him." In his journey, Heisenberg not only learned from Bohr but also understood when it was time to forge his own path. He created a new world of possibilities while remaining faithful to the ethics of research instilled by his "master" and to his initial teachings. He did not rebel meaninglessly against them, nor did he start with the delusional premise of being able to rewrite a field of knowledge by ignoring its foundations; rather, he learned as long as there was something to learn, and used these same principles to open new gates of progress. But above all, by acting this way, he planted the seeds to become a master himself for future generations of physicists, ready to be "followed just enough" and "killed" at the right moment. In a metaphorical sense, of course.

III - The Dopamine Map

In the previous chapter, we discussed the importance of extracting from a situation what is most relevant in order to make these elements the main focus of our action plans. We also talked about some approaches to implement this and how different schematizations can lead to various potential solutions for the same problem. Therefore, in this chapter, we will begin with a kind of *universal methodology to generate action plans,* combining some of the principles we've covered so far with a bit of "creativity science," that is, what research tells us about how our brains generate ideas. But what exactly do we know about this topic?

The discourse on creativity is indeed complex, as I have specified in another of my books, to the point that it becomes difficult, if not impossible, to extract a clear and unequivocal vision from the scientific literature. However, if we were to extract some truth by the most relevant and reliable studies, we would find that, for example, according to Alice Flaherty, PhD at Massachusetts General Hospital and one of the leading researchers on the brain mechanisms underlying the creative process, the generation of ideas is often associated with the release of a neurotransmitter, *dopamine*. To clarify, dopamine is the substance that the brain secretes when we engage in activities that give us a sense of gratification and (potential) reward. This correlation between dopamine activity in the limbic system and the generation of ideas, after all, is something that many of us have probably already witnessed when we've had "sudden" ideas while engaged in intensely rewarding activities. According to other research, however, like that of Joydeep Bhattacharya, professor of psychology and director of research at Goldsmiths University of London, dopamine seems to be a necessary but not sufficient reason. In fact, in the creative process, a decent state of relaxation also appears to be fundamental, such that the two components of our nervous system, the *sympathetic* (dedicated to alert and activation responses) and the *parasympathetic* one (dedicated to "rest and digest" responses) reach an optimal balance. When this balance is in fact reached, our attention can be more intensely focused on incomplete internal processes, creating a condition in which ideas can surface much more easily; and this could be one of the reasons why so many ideas come to us while we're in one of the 3 "B's": Bed, Bathroom, Bus.

That being said, the following chapter can be consulted in two ways: one is the "default" approach, where you can browse through the concepts presented page by page and try to see if, as usual, they suggest something useful or interesting. But "method number 2" involves using it as a practical "Strategy lab". More specifically, you could proceed as follows:

1. Set a goal. Make it something you truly care about, and if possible, formulate it in a way that makes it, as we discussed with the MSP, precisely measurable and emotionally stimulating. Not *"earn more,"* but *"earn x so I can buy my dream house."* Not *"lose weight,"* but *"lose x pounds so I can look in the mirror and love what I see even more."* This will help keep the goal clear and inject the first "dopamine rush" into the process.

2. List the key aspects and classify them by valence. Try to list a minimum of two negative and two positive factors, and ideally, up to four. Remember to prioritize what, given your knowledge, should represent the "20% that holds 80% of the control and power." Most importantly, remember to weigh these factors based on their "real" weight and power, rather than their psychological or perceived weight.

We also discussed aspects that could be potentially "positive" or "negative," but every situation, needless to say, will also have aspects that are mostly "neutral" in character. For example, elements that favor someone over another based on unpredictable or random criteria; uncertain allies, energies not yet well-defined. All those neutral, dormant forces. And especially the details, the invisible forces, everything that is underestimated or underutilized.

In this case, assuming that these neutral aspects are still "winners" in evaluating what is simultaneously more important and manageable, you could very simply divide the aspect into two "possible" sub-aspects, each representing a scenario: negative would obviously be the one where this force could be used against you, and vice versa, and each possible "map" would represent a different case of analysis.

3. For each key aspect in the "dopamine map," review the "actions" in the chapter related to its nature. If the element is positive, go through the "list of positive actions," and vice versa. Read all the various suggestions under each action at least the first time you approach the technique. Then try to imagine "forcing" each action onto the element before you, even if this involves unusual or absurd relationships or combinations.

So, after working on each element, give yourself a "relaxing break" of at least 5 minutes, during which you can take a walk, make yourself a coffee, have a snack, or something similar. Remember that at this stage you should begin to steer your brain towards a phase of "lower cognitive intensity"; thus, it's crucial to avoid shifting from an active work phase to an "equally demanding leisure activity," such as frantic scrolling on social media. It's better to engage in something physical that allows your mind to rest, yet is enjoyable or interesting at the same time. If you're lucky enough to take a short walk in nature, that would be the "perfect practice" to implement in this context.

4. Write everything down. Do interesting ideas come to you as soon as you start making associations? During the five-minute break? Later on? Note anything interesting that comes to mind and try to figure out which ideas are the most worthwhile in terms of cost-benefit ratio (another element we will explore further). And if nothing decent comes up, try again by intensifying the two key factors for generating ideas, which are dopamine activation and relaxation. For example, could you relax "even more" during the 5 minutes? Could you extend it to 10 or 20 minutes? Could you make the goal even more stimulating?

But let's get straight to the list of actions:

Actions for the "negative aspects"

Recapping, examples of "negative aspects" to consider here are:

- *Potential risks.*
- *People who, intentionally or not, can harm us.*
- *Diseases, "distortions," "malfunctions."*
- *Errors we can make, or tend to make.*
- *Bad practices.*
- *Obstacles, blocks, walls to jump over, avoid, or tear down.*
- *Significant losses of time, money, resources.*
- *The causes or the effects of more than one of these things.*

In front of which we can "apply" one or more of the following actions:

Mathematical Prodigies: Conversion

Conversion, in its power, possesses a distinctly "mathematical" rationale. Imagine you have a debt of 100 euros, which on your bank account might be noted as "-100." Applying a strategy based on "conversion" involves devising methods to change "-100" from a debt into a gain. This could mean, in the case of the debt, simply finding a motivation that, prompted by the "burden" of that debt, inspires creative ideas to bring your finances back to a positive balance.

Conversion, at its core, is the process of "mathematically reversing the sign" of something—turning a negative into a positive. This can be achieved by applying forces or strategies that outweigh the existing "negative" influences or by directly altering the key element that defines the quantity in question. This might involve controlling it, reshaping it, reinterpreting it, or even transforming it entirely.

Sun Tzu, in his often quoted (and frequently misused) "Art of War," heavily emphasized this concept: capturing and claiming the enemy's army is "divine," while its destruction is just "mediocre." In the first case, you turn potential energy into a resource. In the second, you still gain an advantage but waste potential that could have been used differently. Clearly, this strategic line (like all others) isn't applicable to everything you encounter. There are diseases that need curing, obstacles that must be overcome, and problems that must be solved, with no hope of gaining anything from what is "dismantled." However, even in these cases, the affliction from the disease, the presence of the obstacle, or the urgency of the problem can always be converted into strength, vigor, and a drive to act, thus ending up "nonetheless" transforming much of one's negative potential into an unexpected "ally" for the journey.

If we wanted to try applying this "conceptual line," we could start by asking ourselves *how truly and irreparably negative the element we're interested in is*. Because perhaps, simply by overcoming certain biases, we might find we can change, control, or turn a negative into a positive. Or perhaps the unpleasant aspects generated or emitted by the element can be rethought, redirected, or even harnessed to our advantage; much like how we currently harness wind, geothermal energy, solar rays, and other natural forces, which, while potentially destructive in some contexts, can be transformed into valuable resources.

Alternatively, as mentioned a moment ago, the approach can be "purely" strategic and, if the negativity of the element cannot be rethought, you can start with the question: *"What resources, tools, and advantages can be leveraged, created, or acquired precisely from all the negative consequences that have resulted or will result?"* Instead of directly addressing our problematic factor, we can try to create an alternative system, a "remedial strategy" that, just like in the previous example of debt, aims to create balance *despite* the problem under examination.

There is no need, after all, for a scientist to understand that life can be a series of unpleasant events: unexpected defeats, people who suddenly leave us, and bad news that *"hits you suddenly on a lazy Tuesday afternoon."* However, the more we instinctively adopt the true "finesse" of converting the negative encounters we face, like genuine scientist-strategists, the more we can paradoxically transform our existence into a continuous, beautiful path of growth, construction, and enrichment.

Playing with Entropy: Weaken

Let's set aside the "subtle finesse" of conversion for a moment and admit that the best possible solution might instead be the more straightforward approach of *weakening, consuming, wearing down*, or even *canceling out* our "enemy force." Perhaps because there's little that can be converted or salvaged within it. Or maybe this wearing down is precisely the path that leads to conversion itself, as in certain medieval battles where the strategy was to exhaust a city through siege until it capitulated, allowing for the looting of its riches.

Assuming, therefore, that as good scientist-strategists we have taken into account that this is a good approach to adopt (and, if possible, that plundering cities is not part of the plan), the questions to start asking would then focus on how to tire, distract, exhaust, disarm, confuse, and render ineffective, irrelevant, or defenseless what interests us.

As we will further explore later, strategies aimed at weakening typically rely on *exploiting the physical principle of a system's entropy.*

Providing a rigorous definition of entropy in this text would be quite complex, especially given that there are two distinct definitions depending on whether one is dealing with thermodynamics or information theory. However, to simplify, we could try to describe it as a *measure of the amount of disorder in any system;* something that, in an "isolated" system, tends to increase irreversibly over time. This is why, for example, you can't piece back together a wooden stick (an ordered system with low entropy) from the ashes (a disordered system with high entropy) into which it has transformed after burning; or, to quote Woody Allen, it's the reason why "you can't put toothpaste back into the tube."

The parallel here with a besieged city in medieval times is obvious: although it represents a clear simplification, it is quite clear that the progressive loss of order and structure is a natural characteristic of certain systems, especially if they are unable to exchange energy or resources with the systems that surround them. Therefore, the approach suggested here could be summed up as aiming to:

- Assign to our "negative forces" those "irreversible transformations" towards high entropy, leading to a progressive loss of structure and energy (imagine the case of destroying an opponent's weapon).

- Put them in a sort of "isolation" and not give them enough time or opportunity to "absorb from external systems" those structures, energies, or resources that might allow them to lower their level of entropy (consider the case of destroying an opponent's weapon and forcing them to surrender before they can pick up another one from the ground).

Working in this way with principles of entropy is simply what has been done, unconsciously, for centuries in every application of the art of warfare. Consider the first piece of advice given in martial arts when faced with someone exceedingly dangerous, such as an armed man or a group of attackers; if you can't flee or negotiate, the first thing to do is to irreversibly *increase the entropy of the "attacker" system,* by destroying their weapon or reducing their numbers, so to improve your chances of survival. Or think of the countless tactics of "preventive weakening" of enemy armies, like those adopted by Mao Zedong, who, also mindful of the teachings of Sun Tzu's "The Art of War," often summarized his military strategy in a formula of sixteen characters: *"The enemy attacks, we retreat. The enemy rests, we harass them. The enemy is exhausted, we fight. The enemy retreats, we pursue them."* This was indeed his main guerrilla tactic against Chiang Kai-Shek: where Mao had far fewer soldiers, he still managed to outsmart every possible way, leading the enemy army into fields unfavorable to them, thereby "weakening" them until achieving victory (obviously, no one here is encouraging you to take any bloodthirsty dictator as an example to follow; consider it merely a case study).

"When the enemy is rested, you must be able to tire him; when he is well-fed, make him starve; when he is relaxed, compel him to move."
(Sun Tzu)

The Quest for Singularity: Restructure

This is a type of intellectual approach that can be attempted whenever a direct confrontation with a "negative aspect" would seem senseless or otherwise inconvenient. Thus, one can ideally try to break down this element in more detail to identify the 20% responsible for 80% of useful effects. And this could be implemented with the "classic" questions about what in the element represents an essential core of the effects that interest us.

The strategy of "restructuring" encourages adopting several interesting principles. The first is the well-known "divide and conquer" attributed by some to Philip of Macedon and others to Louis XI. This concept, the premise of the previous chapter, is used in various forms and colors in scientific research: the more you ideologically fragment something into its components, the more you might be able to channel its dynamics to your advantage. Second, the importance of using this approach to verify the presence of possible "singularities," which we will also discuss in a few chapters. Given the inevitably discontinuous nature of many natural phenomena, it is possible that, especially where complexity increases, a more detailed analysis of their patterns and structures may reveal an aspect that's unexpected, surprising, and not adhering to previously analyzed patterns. Often, the intellectual effort to not surrender to what appears or what we expect from something can be enough to grasp these singularities and derive extraordinarily interesting lines of action. Furthermore, as shown by facts, sometimes merely one extra hour of research, the experimentation with a single new reagent, or the use of a slightly different instrument can lead to discovering a drug that saves millions of lives. The key is believing in the *potential of a tiny step*, as sometimes revolutionary breakthroughs are a matter of millimeters, milliseconds; of infinitesimal differences that can change the entire trajectory of a field of study, an industry, or an entire society.

Like with a cold: Avoid

Whenever a form of intervention on your "element" is not feasible, you can aim to seek possible alternatives to direct confrontation with it. Consider, for instance, the fact that, to this day, there is still no cure for the common cold. The main issue in finding a cure for the cold lies in the huge number of rhinovirus strains in circulation (about 160), which is why it's not about finding a vaccine or a specific cure but, as suggested by Peter Barlow, an immunologist at Napier University of Edinburgh, it is about finding the "master key" to open 160 entirely different locks. Therefore, the problem also involves developing a solution with potentially enormous costs, considering the theoretically minimal outcome of eradicating an illness with relatively negligible consequences.

This example also demonstrates that there's no reason to always view avoidance as a lesser strategy. It might be in fact easy to see it as a withdrawal from more "interventionist" actions, and thus less effective or decisive. However, by now we should have understood that being "scientific" also means avoiding easy judgments dictated by common sense; in such cases, it all comes down to the simple cost-benefit assessment and the realization that it's truly unnecessary to develop 160 different vaccines just to save a few days of a runny nose. Moreover, and we will also examine this when discussing "Playing with Effects," there are analgesics, anti-inflammatories, and that marvelous range of medications that, while not addressing the root of the problem, offer an alternative and extraordinarily effective approach to the issue.

Actions for the "Positive Aspects"

Let's now try to understand what to do in the presence of "beneficial" aspects, which, to recap, might be:

- *Potential opportunities and income prospects, or resource savings.*
- *Allies: people who, deliberately or not, can help us, advise us, and provide us with resources.*
- *Potentially effective strategies and "already existing" best practices.*
- *Strengths "inherent" to our ways of thinking and acting.*
- *Good ideas.*
- *Tools equipped with particular flexibility, utility, value.*
- *Concrete sources of good knowledge and valuable resources.*
- *The causes, or the effects of some above the previous things.*

Forget the Ancestors: Embrace

Sometimes the positive aspect, the good practice, the source of resources, is "already there," and all we have to do is take it and use it. Sometimes using the tools around us to serve our purpose can be straightforward, but in most of cases it is not always easy to catch and understand the potential of the useful forces around us. In fact, the following may arise:

Psychological limitations: Many of our survival mechanisms, being the same ones that governed the actions of our ancestors in ancient times, can produce extremely "curious" effects in us;

consider our natural, evolutionary "love" for a certain type of drama, once a fundamental mechanism for avoiding deadly dangers, or the compulsive pursuit of unnecessary challenges, historically essential to prove one's dominance within the group. All these biases could easily keep us away, without any practical or concrete reason, from the "positive force" in question. We might end up considering it "too good to be true" or "too simple to be an engaging challenge." So, "high" once again on our biases, we might decide to ignore it and proceed in unnecessarily complicated ways.

Once again, the advice in the face of evident psychological limits remains unchanged: self-awareness, attempting to recognize one's own biases, trying to "escape" emotional reasoning, and striving as much as possible to quantify the "pure" benefits of each choice. The more we incorporate these mental "best practices," the more we can "suddenly" access resources and wealth that we might not have been able to consider before.

Concrete limitations: The aspect we're discussing may be "there," but that doesn't mean it's entirely at our disposal. It could be behind a wall, at the end of a road, on the other side of the world, or locked within the heart of someone we love. In this case, the strategic lines to adopt might be very similar to what was seen in the action line of restructuring against negative forces. In simpler terms, this may involve separately addressing the issue of "how and what to use" to overcome the obstacle. It is, therefore, a process that includes identifying the blocks, limitations, necessary steps, and possible allies who can help us make the desired "resource" accessible.

Nonlinear Phenomena: Add and Develop

There are many examples in nature of elements, resources, and structures that might seem inert and incapable of evolution on their own, but when "nurtured and supported by the right allies," they appear to "suddenly" transform into something of unexpected strength and impact. A seed, for instance, is just a small cluster of a few grams of proteins and fibers, but if you provide it with water and soil, it can grow into a tree that produces hundreds and hundreds of fruits. A piece of copper is "just a shiny red stone," but with a tool to draw it into wires, that stone can become a medium for transmitting even very complex information.

All of this obviously should not come as a surprise, as it is essentially "obvious" that if an element evolves in organization and structure, it will reveal an "impact capacity on reality" that might not have been within the initial expectations of an external observer. However, what I find interesting to focus on is, once again, the way in which we become literally unable to perceive this potential.

Consider our natural inability to mentally process time (and how, depending on our mental state, our perception of it can greatly accelerate or decelerate), as well as the evolution of things. Not only do we tend to assign a static and unchanging value to these phenomena, but we also face the challenge of imagining the progressive transformations that can occur over time.

Furthermore, a fundamental limitation of ours is in our tendency to *view all phenomena as linear;* we believe that there is always a constant and predictable progression from point A to point B. However, it has been observed that growth in nature is almost never linear, but rather, it is often exponential, "sudden": a cocoon remains as such for days before suddenly becoming a butterfly. There was no life on Earth for millions of years, and in recent decades we have essentially become capable of creating new life in the laboratory.

From these biological limitations, we can nonetheless derive a set of extremely encouraging courses of action, which can be summarized as follows: always try to grow, always try to develop, always try to achieve more. This is because even the apparent inertia of an element that seems incapable of growing, or at least not as expected, could simply be "broken." We tend to instinctively rely on an altered perception of time, underestimate slow but steady progress, and overlook the explosive potential of exponential change. This means that when it seems a situation or project is stuck or at a standstill, it might simply be that we are in that "quiet" phase of preparation that precedes a sudden and radical transformation, much like a cocoon transforming into a butterfly.

At this point, however, someone might rightly ask: "But how can we tell if this potential for transformation exists?" The first answer I would feel compelled to give is empirical in nature: if you exhaust your time and resources before the desired growth has occurred, then it's likely that no useful potential existed. But considering that this answer might not suffice for you, let's take a moment to look at the very history of our planet.

By observing the emergence of life from a "mass of chemical elements," one might conclude that where there is *complexity, variety, the ability to grow through interactions with external systems, freedom of internal interaction, and rules that encourage the emergence of new stable structures*, there might also be potential for growth. Moreover, even stepping away from natural sciences, it can be confidently stated that there is more potential in a society that promotes free initiative and association among individuals compared to one where corruption and insufficient bureaucracy "kill" each of these phenomena. There's more potential in a mind willing to learn than in one that rejects any form of post-school education, thus succumbing to so-called "functional illiteracy." And so on.

Clearly, the evolution of certain phenomena remains unpredictable, and therefore this rule serves more as a guideline of common sense than anything else. The key, as always, will be to adapt it through careful study of the context in question and, most importantly, through active experimentation of our initial hypotheses.

Think Darwinian: Make it Flexible

"Making it Flexible" refers to fostering and enhancing dynamism, adaptability, and the capacity to adjust. This approach allows the 'positive element' to not only demonstrate resilience but also amplify its positive impact on a broader scale.

Flexibility can represent the most critical determinant within a strategic context. As mentioned before, any "force" that can be extraordinarily disruptive and effective, but cannot adapt to evolving events and circumstances, will remain mostly unfruitful. You can have all the best resources in the world, specifically ready for plan "A," but if "B" happens, they may all end up being wasted. It's precisely this principle that underpins Darwin's theory of evolution: the species that survives is the one most capable of adapting to its environment. *Not the strongest, most agile, or intelligent,* as is often misunderstood, but the most adaptive within the given context. From this, we could gain some added value:

- **By studying the potential scenarios** that may emerge on the horizon. So, you can adapt your resource to "survive," "grow," and "enhance itself"—or at the very least, ensure it generates benefits regardless of the scenario, even the worst-case ones. Doing this doesn't require a superpower or the ability to predict the future. It simply calls for a bit of imagination to envision the most tangible or damaging possibilities. For example, if there's a risk of a significant economic crisis, you might prepare a "contingency plan" to keep your business afloat during tough times. We will delve deeper into this concept in Chapter VI.

- **By enhancing the dynamism and flexibility of your "allied forces"** in all the ways it can be applicable. This could mean spatial flexibility, achieved by reducing their relocation costs. Or structural flexibility, for example, by removing unnecessary "fixed points" of the considered element, thus increasing its ability to change and adapt to new places, contexts, and challenges. This was one of the cornerstones of Napoleon Bonaparte's military strategy. In some of his battles, Napoleon applied a "principle of division," formulated before him by Count De Guibert, through which his army was divided into completely independent sub-units, making them less reliant on geographical constraints and far more "reorganizable" as circumstances changed. Despite the practical division of forces, he achieved a much more effective system, far more ready to execute the orders.

> *"In preparing for battle I have always found that plans are useless, but planning is indispensable."*
> **(Dwight Eisenhower)**

Understanding Life Cycles: Regenerate

We observed this when we introduced the concept of entropy: every force, every element, every entity inevitably ages, wears out, deteriorates, and follows its own life cycle—a progressive "loss" of order and structure. However, since this does not necessarily imply that every individual transformation is completely inevitable or irreversible, it is always worth considering whether it might be possible to avoid, prevent, or at least delay this consumption, this process of deterioration. Moreover, where there is at least partial reversibility, we might even "recharge" and "give new life" to the element of interest.

If it's true that the core principle behind the "science of winning" is that achieving any goal requires a foundation in real and factual principles (or at least ones that seem probable), then it's essential to analyze and respect the life cycles of everything. Moreover, it might be helpful to smartly use these cycles and understand that, while isolated systems inevitably tend to increase in entropy, it's often true that *by stepping back and viewing something as part of a larger system, we can gain a completely different perspective on its life cycle*. For example, if a relationship's dynamics are exhausted, it doesn't mean that completely new rhythms, ways, and rituals can't unexpectedly renew warmth and passion. What was exhausted within that "closed" system of habits and ideas can extend its life by "opening" up to a completely new system. Alternatively, if that is not feasible, one can always try to see if part of what is "destined to end" can be used as a "seed" to give birth to an entirely new system. Consider a failed start-up project, inherently irrecoverable, yet its code can be salvaged to create a new product. Often, just a little more wisdom and "experimental boldness" is enough, and we can try to regenerate even the most unlikely situations, transforming them into fertile ground for new opportunities.

Exchange of Energies and Resources: Maximizing Yield

Natural systems are perpetually engaged in exchange processes, typically related to energy or mass. An intuitive example is a radiator releasing heat into the environment, or a stream of water flowing from a waterfall to a brook below. However, what I find extremely fascinating is that these exchanges can also be modeled in much more complex and abstract contexts.

For example, the laws governing the free market are based on exchange principles, and it's no coincidence that many economic theories take inspiration from natural phenomena. Similarly, interpersonal dynamics can also be interpreted in terms of exchange: in a romantic relationship, for example, time is exchanged for attention, or affection for equal affection.

The intellectual approach of "maximizing yield," in essence, involves *focusing on exchanges* and, more specifically, on what our positive aspects "demand" from us in terms of costs. Money that arrives every month is fine, but in exchange for how much work? Attention we receive from our partner, but in exchange for how much mutual attention? Mechanical energy generated by our car, but in exchange for how much annual fuel expenditure?

Is the return we get in relation to what we give, in short, *acceptable?* Or *can it be improved?* But also, can the *compromise* offered by the exchange be completely replaced with something more "acceptable", given our context or resources? For example, switching from a car to a bicycle can be an option when we are no longer willing to pay too much for gasoline but are willing to invest more of our time and physical effort.

Because while in simple phenomena such as the heat exchange between a radiator and the surrounding air, the efficiency limit of an exchange is well-defined, in a more complex system there might exist latent and potential resources that we have not yet discovered. This allows us to explore new techniques, strategies, and intellectual approaches allowing us to achieve much more with much less.

This can be achieved through simple lines of thought and action: rejecting the current state of things and applying new compromises, new technologies, new solutions, and creative ideas *that make exchanges more advantageous.* Perhaps by rethinking them or rethinking the circumstances in which they occur; by making better use or using resources with greater common sense; by cutting possible "wastes" or unnecessary "dissipations." But also by completely changing the context under consideration, thus *designing exchanges on a level that can generate even greater benefits in return for a much lower cost.*

A technological approach that involves modeling systems as "sets of exchanges" and focuses on completely rethinking them is TRIZ, a heuristic method for solving technological problems developed in Russia starting from 1946 by Genrich Saulovich Altshuller. TRIZ research has shown that many innovative inventions can be developed precisely by "overcoming" compromises between the forces and quantities at play. For example, if a prototype cell phone model is equipped with a hyper-performing graphics card for playing video games in ultra-high definition, but as a result consumes too much battery, the TRIZ methodology would attempt to "maximize the output" between the two by devising a graphics card that doesn't burden the battery at all, perhaps through an autonomous power system.

But even the history of human rights can also be seen as a process of "refining compromises"—a progression from periods where disadvantaged minorities *settled for minimal gains* to moments where they recognized that the rewards for their commitment, work, and efforts *needed to be significantly greater.* And all these examples demonstrate the true "power" of the art of maximizing returns: it's not just about studying how to get more with less, but offering the key to overcoming limitations that are often mistakenly presented as structural, but can in reality be rethought, circumvented, and even completely rewritten.

The Hardest Step: Severing the Alliance

Sometimes the best thing we can do with an allied force is… *sever the alliance*, thereby creating a "favorable opening" and freeing up resources, space, time; thus giving ourselves the possibility to maybe replace the "old" positive force with something new that might bring even more benefits, for a longer time and at a lower cost.

Relationships, people, jobs, habits, energies, compromises, practical means. Our lives often end up becoming crowded with "potential allies" who simply limit us. We stay attached to them out of affection, dependency, fear, or laziness (bias); but in reality, we don't really need them, or worse, their presence ends up harming us, numbing us. If we are "slaves" to this dynamic as individuals, the community we are part of is no exception: just think about how certain dogmatic beliefs, like Aristotelian ones, have been obstacles to scientific progress. Abandoning these beliefs was tough because they were *convenient*, provided easy solutions, and gave us a clear place in the universe, exemplified by *anthropocentrism*. However, it still warrants reflection, in my opinion, on how the scientific approach, while dismantling any notion of an anthropocentric universe, simultaneously provided humanity with the technological foundations for space exploration, life extension, and other extraordinary achievements. In this, it seems to say: *"I don't need to claim you are the most important thing in the universe; instead, I aim to give you the means to become it."* This is where its fierce magnificence lies; perhaps a somewhat poetic image, but *one not far from reality*.

But returning to our main point: severing the inevitable dependency we have with certain "allies" might be the only way to take that extra step, gain that additional advantage, achieve that further goal. Here, it goes without saying, 90% of the effort lies in the ability to discern when an ally causes more harm than good, to admit it to ourselves, and to have the courage to break free from them. Affection, habit, fear, and laziness can be formidable enemies to overcome, as well as the addiction to "the lesser evil" and the consequent incapacity to even imagine the possibility of a better condition. However, a balanced combination of analysis and the spiritual and imaginative courage needed to *envision a better condition* can help us avoid similar pitfalls. And so, it can push us to finally abandon anything that, while masquerading as an ally, is actually pulling us toward our own personal abyss.

"Universal" Strategic Guidelines

Uncover the master key: Trace back to the source

Both in scientific research and technological development, the focus often extends beyond simply creating a result or solving a problem to *tracing back to its source* when it offers valuable insights. This tendency to address causes and origins allows for the development of solutions that can effectively resolve not just isolated issues but entire sets of related problems. In theory, this approach also fosters greater stability and helps reduce potential future costs by preventing the recurrence of similar challenges. In everyday situations after all, we aim to resolve the underlying social or family issues causing disputes, not just quell them temporarily. We don't just produce wealth *"once,"* but ideally strive to create and nurture systems that consistently provide it to us.

Therefore, when the cost-benefit analysis leads us in this direction, here are some questions to consider to apply this course of action (and again, remember that if you feel these lists of questions disrupt the flow of your reading, feel free to *skim through them quickly and revisit them later*):

- What are the main causes or factors that contribute to the creation or perpetuation of the problem or situation under examination?

- Is there a fundamental structural basis or a set of circumstances on which the problem/situation rests?

- Where do the power and influence of this problem/situation come from? Are there specific leverage points or vulnerabilities?

- Is it possible to directly intervene on these causal factors to mitigate or eliminate the problem? Or, if the situation is positive, are there ways to strengthen, encourage, or replicate these causal factors?

- If I cannot directly address the source of the problem or situation, are there other methods, tools, practices, or approaches I can use to replicate the beneficial effects or mitigate the negative effects?
- In the event of a persistent problem, can I develop a solution that neutralizes its negative effects rather than directly eliminating the cause?
- If the element in question is positive, but I cannot directly influence its cause, how can I create conditions that encourage its replication?
- What key elements make this situation positive and how can I incorporate them into a system or process to replicate these results?
- If a certain situation is favorable, can I identify a model or pattern to use as a basis for building a system that reproduces the same effects?
- What resources, skills, or conditions are necessary to replicate or enhance this positive situation in other contexts or on a larger scale?

A fascinating story centered around the creation of systems that "trace back to the source" is probably that of the Bletchley Park group during World War II. For those unfamiliar with it: just before the war, Axis powers exchanged coded messages by encrypting them (thus preventing interpretation by anyone without the same equipment) using a machine called Enigma. Given its indecipherability and ease of use, Enigma was widely employed for both broad and narrow communication. Enigma was actually deciphered once in 1932 by a group of Polish mathematicians (Marian Rejewski, Jerzy Różycki, and Henryk Zygalski), but when the Nazis enhanced its mechanical complexity by increasing the number of rotors from three to five, they made it once again impossible to easily comprehend German messages.

In short, there was a "problem/objective" involving messages to decipher and a clear "source" for all of this, laying in the mathematical complexity of the machine generating the encryption. Therefore, given the situation, it was clear that the necessity was not so much to work on *deciphering the messages individually*, but to *trace back to the source* and "defeat" Enigma; in this case, by building something complex enough to decipher the German messages.

And it was here that the work of the Bletchley Park group came into play, the main center for cryptographic information analysis that worked tirelessly in the United Kingdom to decipher Nazi messages. In this context, we can't forget about the work of Alan Turing, who later died by suicide in 1954 due to persecution by the British government because of his homosexuality.

Turing and the Bletchley Park team built the *Bombe*, a machine that automated deciphering Enigma messages, thus defeating Enigma. Thus, Turing and his team not only laid the groundwork for the development of extremely advanced cryptanalysis techniques for the time, but also provided an incredible amount of strategic information to the Allied forces. Although the "weight" of this work is still debated, it seems to be the common opinion among historians that the information from Bletchley Park's efforts significantly shortened the duration of the war, saved countless lives, and contributed to the gradual dismantling of the Nazi war machine.

Predators and Prey: "Playing" with the Effects

If you can't, or it's absolutely not advisable to intervene on the considered aspect or its source cause, you could still try to "play with its effects." It's the example we brought up a few pages back when we talked about the advantage of using symptomatic rather than curative drugs for certain conditions. Or it's the dynamic where, when caught in the middle of an economic crisis, an individual or organization doesn't try to address the causes of the crisis but instead seeks to develop entrepreneurial, personal, or project-related ideas that allow them to "navigate" and survive the "lean times."

When deciding to "play with the effects" of a "negative" force, in short, you can base at least part of your strategy on:

- *Avoid the occurrence* of such effects.
- *Prevent* such effects. Work to stop their generation or at least their "impact."
- *Protect yourself adequately* in case they occur.
- *Channel and redirect these negative aspects elsewhere.* Alter their "target" to make them less harmful, inconsequential, or even empowering, beneficial, and advantageous.
- *Simply let them happen* and try to "heal the wounds" and "strike back" after the fact has occurred.

While considering a positive aspect instead, we might consider actions such as:

- *Multiply and amplify* the effects of our appearance through tools, techniques, and additional resources that allow it.
- *Redirect* them where their impact is multiplied.
- *Focus* them in such a way that resources and energies, which "normally" impact different aspects, are entirely redirected towards a narrow set of them, thereby multiplying their effects.
- *Simply let them manifest* and "impact" what interests us, then act on "enhancing" and "expanding" the effects of such impact further.

When we consider natural dynamics, this occurs in a myriad of situations. Prey animals often cannot compete with predators in terms of strength or speed, so they develop self-defense mechanisms that do not rely on "force against force," but instead use different strategies such as camouflage (think of butterflies or certain lizards), or secreting poisons or foul-smelling liquids (think of certain species of toads or skunks). In the market, very similar dynamics can be found: many start-up companies fail early in their attempt to compete with major corporations on similar products. Other companies, on the other hand, recognizing the reasonable impossibility of entering certain market segments, aim to "play with effects," "counterbalancing downstream" the negative effects of their competitors, and creating completely new market niches from which to derive value (the so-called "Blue Ocean Strategy").

After the relative flop of the GameCube, for example, Nintendo decided it was time to change its market strategy compared to its bigger competitors, Sony and Microsoft. Realizing it was difficult to beat them at their own game (ever more powerful consoles with increasingly realistic killer applications), the Japanese company behind Super Mario decided to "play with the effects" of what Sony and Microsoft represented, rather than "attacking them directly." So they shifted their focus to creating a gaming console that wasn't very powerful but was "social," designed to be used live with family and friends, targeting not the hardcore gamer market but the casual one. This led to the development of the Nintendo Wii, which by 2020 became the fifth best-selling console in history (just a few units behind the fourth, the original Sony PlayStation) and a machine that completely outperformed its competitors of that generation.

"If you can't fight reality, build a model that makes reality obsolete."
(Buckminster Fuller)

Strategy Lab - Costs/Benefits

As you may have noticed, in most cases, any creative idea processing procedure does not lead to a clear line of action to follow from point A to point B but typically results in the identification of many possible options to act upon, and consequently, many possible strategies.

The reality of limited resources inevitably forces us to make deliberate choices about the direction to take. This demands a process that helps us overcome the cognitive biases our minds are prone to and focus on decisions that succeed based on the often-cited "cost-benefit ratio." In other words, choices that effectively answer the question: "What allows us to achieve more with less?" Given the critical importance of this principle, let's explore a possible implementation of it in detail:

- For each possible strategy, write a single number corresponding to its "cost" in terms of resources, energy, and time. If you don't already have a precise quantity to work with, try to assign that option a rating or estimate that corresponds to its "cost," "consumption," or "harmfulness." It can be a rating from 1 to 5, 1 to 10, or 1 to 100. The important thing is that the criterion and scale used are the same for each item.

- If you still feel uncertain about how to assign a score to a choice, list all the problems, costs, and disadvantages that come to mind for that choice. Then, assign a "damage score" to each of them and calculate the mathematical average.

- If any of the indices written relate to a problem, a cost, or a disadvantage that is uncertain, try multiplying each of them by a "probability coefficient." Multiply them by a number ranging from 0 (it won't happen) to 1 (it's certain). Clearly, using "0" doesn't make much sense since it would nullify any amount and wouldn't make listing that specific cost very useful. However, if you'd like to avoid this, in cases of very improbable events, you can use very small fractions like 0.1, 0.001, and so on.
To estimate probabilities for everything else, ask yourself: "How many times might this occurrence happen out of a hundred attempts?" Then divide the resulting percentage by 100 to get your probability coefficient. For example, in the case of an uncertain event like a coin toss, the probability would be 50%, divided by 100 = 0.5. In the case of a nearly certain event, 90%, divided by 100 = 0.9, and so on.

- Here too, you might have a definite or almost definite probability distribution from your data regarding the occurrence of something (for example, you know you'll win with a 50% probability, or 0.5), or you might have a simple estimation. The important thing is to be aware that the less accurate the probability measurement, the more you risk straying from conclusions that can accurately reflect the reality of the situation. If something still isn't clear to you, set this point aside, and we'll delve deeper into the theory of probability in a few chapters.

- If you're still here instead, replicate the steps illustrated but take into account a "profit index" or "benefit" for each of the options considered.
- For each option, use your calculator, phone, pen and paper, or whatever you prefer, and divide the benefit index by the cost index.
- "Pruning": The best strategies will be those with the highest index. You might realize the benefits of applying quantitative criteria to a decision-making process: the number calculated here could reveal that the solution that seemed most enticing is less constructive than initially appeared, or vice versa. And that's where the magic of measurements lies!

Strategy Lab - Project Pillars

A few pages ago, we discussed the importance of the strategy of being "flexible" and how, in certain contexts, this is a key factor not only for survival but also for the optimal management of complex situations. Consider, for example, biology, where adaptation is a fundamental part of the dynamics that govern a species' ability to perpetuate itself. Similarly, in the free market, a company that effectively adapts to various changes is much more likely than others to maintain its market share in the long term. We also find similar concepts in cybernetics (the science that studies complex systems), such as the "law of requisite variety". According to this law, one system can control another only when the first is able to achieve a greater variety of states.

Let's focus for now on the importance of the ability to "express variety." Here, while variety as an intrinsic quality can indeed be an advantage, it's also true that applying such variety through excessive flexibility can present challenges. Consider the CEO of a company who constantly changes their mind about the business plan without showing firmness in their direction: they would likely end up wasting time and resources, potentially leading the company to failure. It should not be difficult to understand that without a certain degree of stability, without a minimum of structure, without the persistence of certain forces in one area, it is also very difficult to achieve a level of strength, energy, and complexity necessary to enact significant change.

> "One must always have an open mind, but not so open that the brain falls to the ground."
> **(James Randi)**

Once we've identified our objective or problem, another effective way to generate creative and strategic inputs is by defining cues that help strike the right balance between variability and stability. This means identifying and establishing precise pillars that define the concepts from which we cannot deviate, adopting the philosophy of questioning everything that doesn't belong to them. And here is an example of a working method we could implement to leverage this principle:

Identify and outline the primary fixed elements of your project: start with the facts and objective realities, then proceed with the components that definitely work in the chosen field and all other unchangeable constraints imposed by the context and the project itself. These can be physical constraints, time constraints, or resource constraints. Additionally, they may include more abstract limits, such as the set of needs, goals, and ideals that this objective should fulfill. For example, if it involves developing a new type of pencil for writing on tablet screens without any lag, you could define "primary fixed elements" as the usable materials, the maximum input delay in milliseconds required by the object, the budget available, the technologies with which similar projects have already been completed, and potential investment sources.

Ask yourself if the primary constants are sufficiently essential. Too many constants, or constants defined too specifically, can be an obstacle in terms of "necessary variability" and flexibility. Review everything you've written in the previous section and ask yourself: "Is this necessary?," "Is it really a constraint?" and "Can I define the same thing with fewer terms?" Perhaps you are imposing unnecessary restrictions on yourself, or you are pointlessly aiming for something too detailed from the start, instead of reaching it gradually.

Examine possible illogical aspects and contradictions. If there are constraints that conflict with each other or with those implicitly set by reality, you probably need to take action, and you can do this in two different ways: 1) Redefining them to the point where they can coexist, or 2) Eliminating the "weaker" ones. For instance, if within the same project you want to achieve high sales volumes while targeting a very tiny niche, you might be dealing with contradictory constraints. In that case, how could you "resolve the illogical aspect"? By expanding the niche? By reducing your profit expectations?

Find relationships. Can any of the listed pillars reveal relationships that might be useful to you? New cause-and-effect relationships? Schemes? Patterns? Recurrences? New information on what might "weigh" more within the context you're interested in? Note everything and see if these relationships, in themselves, might represent new primary fixities.

Now persist with your primary fixations. As for the rest, you have total freedom. Therefore, as a first step, try to define your next concrete move. In fact, in attempting to decide what that might be, you may immediately realize that you have "too much" freedom, to the point of being unable to choose. If that happens, simply try going back to the first point: it could be that you're in a situation with a "lack" of constraints, and thus, you need to add new ones to help you better focus your actions.

It's also possible that the initial draft of primary fixations has already revealed some interesting ideas about the possible directions to take, but it's simply unclear which ones are worth starting with; and in these cases, you can simply impose some "priority criteria" on yourself: is it better to do the most important things first? The most urgent ones? Those more dependent on the exhaustion of time or resources?

Alternatively, a different approach to the work method we discussed could be the adoption of the so-called "secondary fixities." These represent support constraints for the main pillars and are subordinate to them. However, such secondary constraints can only be considered as such if they meet three conditions: 1) They are effective, 2) They do not exceed a predetermined limit of resources, time, or expense, 3) No better alternatives emerge that offer a more favorable cost-benefit ratio in the long term.

Let's reconsider the example of our "magic tablet pencil." The "primary" constraint is located in the process of sourcing materials for its creation. It might happen that, after an initial phase of research, you decide to establish a secondary constraint: to use a particular copper-tin conductive alloy, which seems to offer the best cost-benefit ratio. Your secondary constraint might then involve dedicating the next three weeks to exploring how to implement this technology. This constraint will remain in place until you discover, for instance, that the outcome is not as satisfactory as expected, and here the fundamental key will be to keep focus on the primary constraint, never relinquishing it; something that we tend to do too often is the proverbial "throwing the baby out with the bathwater." Something contextual, specific, and isolated doesn't work, and yet we tend to question what remains as well. In this view, reasoning through primary and secondary pillars (or even tertiary or more, if we want to establish more complex hierarchies) allows us to avoid this trap, enabling us to stay focused on "discarding only what doesn't work and keeping everything else."

"If you know you are on the right path, if you have this inner awareness, then nothing and no one can stop you, no matter what they say."
(Barbara McClintock)

Strategy Lab - The Eight Forces

Let's draw inspiration from what was discussed about entropy a few pages ago. In summary: an isolated physical system typically tends to gain entropy and lose order and structure over time. If we're interested in wearing down or destroying a system, it might be advantageous to increase its isolation and accelerate its irreversible transformations. Conversely, if we're interested in making a system grow and develop, it might be beneficial to "reduce" this isolation and allow the addition of tools, techniques, ideas, energies that, where possible, reduce its entropy, thus allowing it to gain in organization, strength, and power.

In this regard, the "technique of the eight forces" is merely another creative technique (which you have probably already encountered in other texts of mine, albeit in different variations) aimed at providing possible inspirations on how to encourage the growth of something to its maximum potential, and perhaps even break through potential "barriers" in its evolution. This likely has countless practical applications, considering that almost any concrete objective can be formulated in terms of "growth of a variable," whether it be an indicator of our economic, physical, or emotional well-being; or maybe the amount of energies and resources accumulated for a specific purpose.

With this in mind, let's immediately look at the "forces" we could apply to our system to make it evolve, here devoid of any true "scientific" classification but differentiated solely through a criterion of "evocativeness of meaning."

- **Ability to Take Risks:** Courage, responsibility in facing the unknown, the capacity to take initiative and assume risks proportional to the difficulty of the set goal. This force can be enhanced, for example, by asking questions like: "What are the risks that I have never taken and why?" "What is the most frightening? What is never faced openly? Has the time come to confront these things?" "What keeps me in my comfort zone the most?"

- **Creativity:** The ability to see or pursue new paths and adopt unconventional solutions. Here you can ask questions like: "Could I look at some elements from new perspectives?" "Should I ignore everything that is common knowledge regarding the situation?" "Should I act in a less predictable way?"

- **Ingenuity:** The ability to employ means, tools, reasoning techniques, advanced technologies, and strategic lines that, by adding complexity and structure, enhance the organization and the final "yield" of every resource. Here some valid questions could be: "Is there any dormant or potential energy that I could harness?" "Can complexity be increased?" "Could I benefit from systems that already perform some of the functions required by my process?"

- **Willingness to Renounce:** The ability to sacrifice what one possesses or takes for granted, and to do so even more if the system demands it to "give more." Here some valid questions could be: "What exactly would I never, ever give up? And to what extent does that represent a real limitation?"

- **Energy:** Meant as physical, mental, thermal, kinetic, electrical, or anything capable of accelerating, within the given system, an evolution towards the chosen goal. The questions to add more "energy" could be: "Where would an extra effort be needed? What kind of effort?" "Should I take one, two, or ten more steps that I normally wouldn't take?" "What stops me from putting in more effort? Laziness? Fear? Lack of energy or focus? How could I deal with these things more productively?"

- **Organization:** What kind of structure can be given to the elements at play? How can they be combined? What can enable them to merge their potential and energies? What can ensure they perform at their best? Are there methods, tools, techniques that can create a better synergy among the elements at play?

- **Time:** Willingness to dedicate one's time to the set objective, ability to optimize processes, and to allocate an appropriate amount of time to each of them. Here some interesting questions to ask may be: "Does every element truly receive the amount of attention it deserves?" "Should I eliminate time-wasting activities?" "Is time being dedicated to "20" factors that could instead be dedicated to "80" factors?"

- **Other resources:** Anything else the system can utilize, whether it's money to add to the budget, reagents to start a combustion, people to include in a team, or nutrients to support certain biochemical processes in our body. Here we could ask ourselves some more generic "What can typically make the system grow? What could represent added value?"

So the secret is: "Take one or more of these things and keep adding until the result appears"? Like anything real, unfortunately, it's not that simple. The first problem is: given the inevitable physical, chemical, informational, and spatial limitations of almost any existing system, one must always be wary of the concept of saturation. Simply put, if one element helps another grow to a certain level, beyond that same level, its contribution typically becomes counterproductive (think of a plant killed by too much water, relationships smothered by too much attention, or a team that no longer produces because there are too many people to coordinate effectively). Therefore, before starting with the technique, also pay attention to the following concepts, as they may prove useful in countless real-life situations:

Remember the Pareto principle: If you want to "get more" and make certain elements within the system "grow," avoid resorting to "brute force" by hastily using your resources and energy in the first way that comes to mind—or worse, in "massive doses." Instead, as discussed a few pages ago, focus on applying your efforts to one, or at most two or three, key aspects that hold the greatest influence and potential. These pivotal elements can "trigger" the maximum result with the least effort and most acceptable cost.

Before increasing, harmonize. Where you find yourself needing to "increase" or "add" something to achieve more, pay attention to the sustainability of such an increase; also, employ everything at your disposal so that this addition can "harmonize" effectively with the existing system. Most modern fitness coaches, for example, will tell you that if you truly want to improve your physical condition, you shouldn't just exhaust yourself at the gym until you can't see straight; rather, you should try approaching your problem from a holistic, global viewpoint that includes motivation, method, and a general improvement of your health and energy. Thanks to this approach you won't just probably get in shape "automatically," but you'll likely be able to handle much higher amounts of intense physical effort more effectively.

Keep in mind all those "biases" that will lead you to underestimate and underutilize the resources at your disposal:

- **"The grass is always greener on the other side":** our focus is too often directed towards other spaces, times, and dimensions; and this is both because ideals and dreams become increasingly alluring and because we tend to amplify the perceived value of anything that demands significant effort. Therefore, start with what you have and always ask yourself if you aren't underusing or underestimating it; maybe, because you have become "accustomed" to it.

- **"Trying to reinvent the wheel":** considering both our obvious informational limits and our "love for effort" mentioned earlier, we too often tend to want to recreate from scratch solutions, structures, or resources that already exist. Before starting from scratch, simply research already invented wheels. Then, evaluate and compare the costs of the two approaches, because it's likely that at least 7 times out of 10, you'll realize that choosing not to recreate is the one with the best cost-benefit ratio.

- **"Relax":** According to an essay published in 1958 by Cyril Parkinson, "Given time to complete a project, we are likely to occupy all of it." It doesn't matter if we are given three or six months to do exactly the same work; it is probable that in any case, we will "stretch" the work to fill all the time we are given. With due limitations and attention to extreme cases (imagine being given two days to build a travel company to Mars) this means that when a quantifiable resource is scarcer, we usually tend to use it better. Therefore, whenever you are dealing with "quantifiable" and "limited" resources in a project, such as your money, available time, or space in your living room, practice an exercise: try to think as if you were forced to work with just a tenth of what you have at your disposal.

- **"Noise and Common Sense."** In simpler terms, an element's degree of usefulness or harm is often judged by how well it aligns with common sense and the emotional response it evokes in those evaluating its value.
Yet, consider how some remarkably useful resources can emerge precisely from what might appear useless or dysfunctional; such as when a hideous illegal building is transformed into an architectural masterpiece through graffiti, or when a broken cup is turned into a work of art using the *kintsugi* technique. But also the way we tend to prefer a more "evocative" or "familiar" solution, rather than the most effective one, as with many people who choose branded drugs over generics, even though they are chemically identical.
This principle, in essence, should remind us once again to rely more frequently on objective measurements and assessments of the advantages or benefits of something, while also theoretically not underestimating details, quieter forces, those that no one

pays attention to, or even those preconceived as "ugly or unpleasant."

Invert. In other words, try to understand if things aren't going well perhaps because you are already in a phase of saturation. Returning to the earlier example: instead of trying to figure out how to do more exercise in the week, simply aim to do less. Once we are free from the pressure of having to do something "at all costs," we might actually find ourselves doing it more spontaneously and more often than we initially planned.

Strategy Lab - The Seven "Muda"

If the previous technique was primarily based on improving a system through adding and harmonizing, the "muda" technique is instead based on the opposite concept of "elimination," specifically focusing on production processes within the system itself. "Muda" (無駄) is a Japanese term that can be translated as "waste" or "non-value-adding activities," and a technique based on identifying and removing these elements was first implemented by Taiichi Ōno, a manager at Toyota. By placing this philosophy at the core of his "lean manufacturing," he quickly transformed the company from a small firm into the world's leading car manufacturer.

Obviously, scientific processes and business processes are governed by philosophies that are at least partially different. The essential "gist" here, however, primarily relates to what we discussed about "maximizing yield": there is generally no process that doesn't benefit from an analysis aimed at maximizing the yield of its participants. Furthermore, much of this also connects to what was said about the importance of maintaining a certain degree of flexibility: systematically removing the non-essential can allow us not only to achieve the maximum with the minimum, but also to be less vulnerable to potential changes.

But let's look at the practical application of this technique, which is actually quite simple: review the "muda" lists below and try to determine whether each of them can be worked at. All of this should be done without being too "ruthless" in its application, and always giving, depending on the context, at least a percentage of leeway to partial redundancy.

Defects. In the manufacturing sector, the time and resources wasted fixing a mistake are rarely comparable to what would be required to prevent it. Obviously, this doesn't apply to every process and, as mentioned, there will be countless practical situations where it's more advantageous to venture into at least partial experimentation, rather than trying to master uncertainty through some form of ultra-control.

Overproduction. That is, the act of creating something redundant; perhaps solely for the purpose of following a belief, an existing practice, a habit, or an obsolete procedure.

Unnecessary waiting. One of the key elements of lean manufacturing is the concept of "flow." In this perspective, everything should move smoothly without delays. Eliminating or using these delays positively, like reading a book while waiting at the doctor's office, can be a useful approach.

Unnecessary movements. Just as with overproduction, how many times is a movement, transport, or relocation carried out not because it's truly needed, but simply out of habit, the need to adhere to "common practice," or a lack of imaginative capability? This is probably one of the "muda" most forcibly applied in the corporate world due to the Covid-19 epidemic: with movements legally restricted to contain the outbreak, many companies were compelled to implement remote work practices, suddenly realizing how numerous activities based on the exchange of information and services did not require constant physical movement to be effectively executed.

Stock. Sometimes also referred to as "warehouse," this muda refers to everything that lies inert and unused, thus not creating any useful value. Consequently, its removal can involve periodically assessing all "unnecessary stock" and everything that is underutilized or underestimated in a given situation; a possible subsequent step is to evaluate the potential cost-benefit ratios of either: 1) Reworking it into something that creates value, or simply, 2) Disposing of it.

Useless processes. This "muda," conceptually very similar to that of "overproduction," emphasizes the importance of simplifying and streamlining processes that fail to add real value to the final outcome. Addressing this waste often involves experimenting with the removal of such non-essential elements to evaluate whether the system, as a whole, remains stable and functional over a reasonable period of time. This approach not only uncovers inefficiencies but also helps in identifying opportunities to improve overall system performance by focusing on what truly matters.

Wasted potential. What has untapped potential? And what biases, what preconceptions, what obsolete systems prevent us from recognizing or harnessing the "clear value" of something?

Stories of "Scientific Champions"

The Silent Odyssey of Maurice Hilleman

Hidden behind the shadow of more prominent names in medicine, Maurice Hilleman is one of the lesser-known figures to the general public, but his impact on global health is unparalleled. We are talking about the man who developed over 40 vaccines during his career, many of which are still used today to prevent devastating diseases.

Born in 1919 in Montana, Hilleman lost his mother to tuberculosis two days after his birth. This tragic event might have sparked an early desire to combat infectious diseases. Raised on a farm, his initial upbringing did not suggest an extraordinary career in medicine, but challenges and adversities can often shape resilient characters.

Hilleman earned his doctorate in microbiology and immunology and began working in the pharmaceutical industry. His career took a turn when, one night, he recognized the symptoms of a mumps epidemic in his son. Collecting a sample from his son's throat, Hilleman set to work and, in a short period, developed a vaccine for mumps, which later became part of the renowned MMR vaccine (measles, mumps, and rubella).

But mumps was not his only success. Maurice Hilleman also developed vaccines for hepatitis A, hepatitis B, chickenpox, meningitis, pneumonia, and many other diseases. His research had a global impact, saving estimates ranging from millions to billions of lives over the years.

What makes Hilleman's story so powerful? It's not just the extraordinary number of vaccines he developed, but also his unwavering dedication and his rapid response capability in the face of healthcare emergencies. For instance, when an Asian flu epidemic threatened the United States in the 1950s, Hilleman swiftly reacted by isolating the virus and helping to produce a vaccine in record time, thereby preventing a large-scale epidemic.

Yet, despite these extraordinary achievements, Hilleman remained an unsung hero for much of his life, and only recently has his legacy begun to receive the attention it deserves. His story serves as a further reminder that true impact often lies behind the scenes, and that those who work quietly, with dedication and a pragmatic spirit, can have a profound and lasting effect on humanity. Which, in my opinion, in a world of "endless noise," takes on an unimaginable value.

IV - Scientific Hacking

So far, we have mostly discussed the art of crafting a "scientific strategy" through different philosophies, each with a unique approach. And so now, while continuing throughout this journey, we will start talking of what might seem like the most "mischievous" approach of all (at least so far): *"How could I "hack this system, here? What set of tricks offers me an unconventional advancement path, a privileged gateway?"*

Obviously when some of us hear the term "hacking," it often brings to mind a strictly IT-related context: hacking as in breaching a computer, a smartphone, or even a bank's security system. However the term "hack" is used in a much broader range of situations: you may have heard of "life hacks," which refer to small practical tricks for everyday life, for example . It's precisely in this sense that I want to conceptually set the stage for this chapter: referring to "hack" as a strategic approach based on *seeking an unconventional solution to our problem*; a method primarily focused on hyper-effectiveness, through identifying a set of *clever tricks, loopholes, shortcuts in the relevant context*.

I may know what you're thinking here. When one starts introducing words like "loopholes" or "shortcuts," it also begins to venture into morally ambiguous territory. Which is *absolutely true:* these approaches could also be used to steal, exploit, manipulate, and harm. That's why I'd like to make some necessary disclaimers: 1) I take for granted that, whatever tool is introduced in this book, whoever reads it is also a "noble scientist," endowed with the basic common sense necessary to use them without harming anyone or doing anything illegal. 2) If giving in to the "dark side" of the "science of winning" ends up ruining your life, your finances, or your relationships, know that it will be solely and entirely your responsibility, and 3) Perhaps the most important of all: **these are just tools!** Yes because, just as an example, we can use what we know about atoms to create terrifying weapons or clean energy, and much of the outcome will depend on the philosophy and methodologies we decide to employ.

Of course, it would be foolish to deny that some tools lend themselves more easily to more "malevolent" uses than others: one certainly cannot claim, for example, that a firearm firing 200 bullets per minute is comparable in potential damage from misuse to a Swiss army knife (and anyone who argues otherwise is either naive or intellectually dishonest). But in my opinion as long as we're not talking about weapons, but rather simple intellectual approaches, we're still in a safe enough zone to confidently say that it all depends on the way they're used.

Never forget that it is precisely thanks to loopholes, "tricks," and the search for vulnerabilities that the first steps of scientific thought became possible—especially in times when criticizing dogmatic knowledge was profoundly risky (does the name "Giordano Bruno" ring a bell? If not, do some Googling...). We could say not only that these tools don't necessarily have to be used improperly, but also that, historically, innumerable circumstances show how a "misguided" conceptual approach can become necessary. This occurs whenever power dynamics with those committing abuses become insurmountable, making it essential to shift the battlefield to a plane where one can effectively defend against aggression, violence, and domination. From this perspective, I find it absolutely invaluable that such tools are added to the intellectual "arsenal" of the reader— because science must embrace the ability to think truly outside the box, question reality in a critical and creative way, and discover profoundly innovative solutions. And very often, this needs to happen despite external, dogmatic, malevolent forces that seek to suppress dissenting ideas or maintain the status quo; for which reason, yes, sometimes we need to be a bit more like "hackers." Let's see more in detail how.

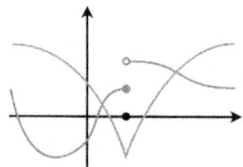

Useful discontinuity

As discussed when we talked about the "Science of Deception," accepting that the things that come to our attention *might be very different from how we imagine them*—such as in their future developments or aspects we do not directly experience—can prove to be an excellent tool for understanding reality.

By adopting this perspective, we can indeed approach the system we're interested in through the principle that even the most powerful and well-defended of them will still have limitations in terms of entropy (as mentioned a few pages ago). Therefore, it might not be equally strong in every aspect, every phase, every behavior. It is in its inevitable exceptions, discontinuities, and inconsistencies that vulnerabilities, openings, and "parts" from which to gain an advantage can potentially be identified.

This also relates to what we mentioned when discussing the "restructuring" strategy: the progress of science is filled with discoveries showing that different natural phenomena, initially believed to be "necessarily harmonious as they are created by God," have actually been found to possess points of discontinuity and non-uniformity. The evolution of species is anything but a natural drive toward perfection and is characterized by numerous halts and branches. The life cycles of celestial bodies are far from linear. The study of the extremely small or the extremely large constantly uncovers new variables that are difficult to integrate into existing models.

It's clear that the "useful discontinuity" of our interest may not even exist. But there's no need for me to tell you that this principle, just like all those that will follow, certainly *doesn't aim to be a universal tool for solving every problem*; the important thing, once again, is to approach the world out there with genuine intellectual curiosity. Explore things from the standpoint of your own ignorance and muster all the willpower needed to take that extra step, to read that extra page, to have that extra conversation, because that's exactly how you might reach that truth as surprising as useful. But let's move on to the questions we might ask ourselves to practically guide our investigation toward "finding the useful singularity." And again: if you think that such questions list disrupt the flow of your reading, skim through them quickly and return to them later.

- Could this system be at least partially different, weaker, more vulnerable than it appears to be?
- Can it also be the opposite of what it seems to be?
- Does its most evident appearance grant it "fake," superficial power, control? Could this appearance have been crafted solely for the purpose of masking a weakness?
- Is what I don't see shaped by my preconceptions and my imaginings?
- Is this thing clearly too large or complex to be identical in every aspect or moment? Am I being deceived by having seen or perceived just a part of it?
- Where and how could one hide points, phases, weak moments?
- Could I have overlooked some details? Am I ignoring the possibility of a limited singularity from which I could still extract the advantage I need?

Moving "sideways"

The term "lateral thinking," which you have likely heard before, was coined by psychologist Edward De Bono. It refers to a problem-solving approach that involves first reimagining the problem, gaining an entirely new and different perspective on it, and then devising a set of solutions based on completely non-obvious or unconventional considerations.

This is a concept that will feed into and form the foundation for many of the considerations that will follow. It's not difficult to imagine that those who aim for a strategy of "scientific hacking" will, in most cases, derive great benefit from devising a line of action that is creative, disorienting, "unexpected"; that's designed to strike right where the system is not expected to "be attacked".

However, pay attention: conventional procedures and ideas might still be the most functional and convenient ones to adopt in a given context. We saw it just a few pages ago, after all: the search for a creative, lateral, unexpected solution can also be extraordinarily fascinating. For this very reason, it is though crucial to remain *rational and pragmatic;* we need to look for what works, not for what's most fascinating. And even a rejection of the conventional must stem from a clear awareness of the limitations to which the existing domain confines us. Therefore, "lateral hacking" serves as a strategic toolset, to be employed only when such limitations tangibly obstruct our pursuit of the renowned goal: achieving the "best result with the least effort."

To summarize these considerations into small, straightforward guidelines, we could say that applying the "principle of laterality" can yield real advantages only when:

- **Obvious problems, difficulties, or limitations** exist within the realm of the common or the known.

- **Sufficient time and resources** are available to invest in uncovering new perspectives and, consequently, exploring lateral courses of action.

It is also worth considering that, even here, our course of action shouldn't be constrained by overly limiting classifications. In most cases in fact, *there won't be purely direct maneuvers and purely lateral maneuvers;* rather, the best solutions may gain strength and impact precisely by combining the two approaches. "Shock" with the unexpected and then "reassure" with the predictable; or "capture" with the predictable and then deliver a "final blow" with the unforeseen. And so on.

That said, let's try to list some basic general rules to attempt "lateral" approaches to the system we are interested in. As usual, these rules do not implicitly provide any answers, but they might at least "open up" the right direction to investigate:

- **Rephrase the problem.** Rewrite it using (almost) synonyms, different constructions, or by emphasizing unconventional or lesser-known perspectives.

- **Research, create, and innovate.** Don't just stick to common procedures to solve a particular problem; use your imagination. For example, try taking a sheet of paper and write down ten potentially absurd ways to solve the problem; then, "push yourself" to understand how they could actually be implemented.

- **Surprise**, anticipate, try to figure out how to be "among the first" to arrive.

- **"Occupy the spaces,"** and don't allow the system to prepare its necessary "defenses" nor to "strike first." Try to plan an initiative even if it goes against opposing signals.

- **Be "irreverent."** Do not accept limits imposed by overly strict morals or archaic and bigoted conventions. Try to devise a strategy that will "inevitably" provoke someone's disapproval (within the boundaries of legality and common sense, of course). Try to be bold.

- **Break the assumptions.** If everyone else takes something for granted, try to challenge this rule (unless it's something in the realm of objectivity or law, of course). If everyone else would limit themselves, hesitate, or confine themselves within a conceptual wall, simply make it your duty to try to overcome it.
- **Go the "extra mile."** If there's a point where you would normally stop, just aim to do more, do better, and reach the "next level."
- **Be brave.** If others would be afraid to do it, do it. If in the past you would have been afraid to do it, do it a hundred times over.
- **Embrace chance.** Provided it doesn't mean ignoring any objective reality and the risk and cost are manageable (see: don't do anything foolish!), try an action that is *random*, that is nonsensical. Something that might even seem completely at odds with any pre-established practice, reasonableness, or common sense.
- **Try to plan more often initiatives where "no one expects them,"** including the system itself or those involved.

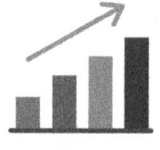

Predictability

So obvious, yet so powerful: an excellent technique for identifying a system's "vulnerabilities" lies in precisely focusing on maximizing the understanding of its behaviors; the more patterns, habits, repetitions, recurrences are recognized, the more likely it is to find "gaps" and flaws that could lead to the kind of advantage we're seeking. It's the cliché example often seen in Hollywood, where protagonists sneak into a building while the night guard falls asleep, or where a recurring vice of the antagonist ultimately reveals to the hero how to defeat them.

Since I am quite the nuisance, it is once again my duty to remind you that our much-loved reality comes with its fair share of "buts": firstly, borrowing from what was said a few pages ago, one *might fall into an inductive fallacy here*. In fact, there will be countless occasions where the phenomenon "A" has occurred even a million times, yet it's not guaranteed that'll happen once again. Previous data might suggest it's *more likely to happen;* however, except in the case of certain hard sciences, predictions must always account for a degree of fallibility. After all, Bertrand Russell's famous example of the "inductive turkey" is timeless: fed daily since birth, only to be slaughtered on Thanksgiving Day, the turkey, even on the eve of its demise, believes it will be fed forever—utterly incapable of foreseeing the sudden shift in the laws that had always defined its "reality."

But it's also crucial to understand that, no matter how predictable or repetitive a pattern may be, this doesn't necessarily grant us greater power over the entity it belongs to. For instance, we can precisely calculate at any moment the angle of rotation and the relative position of the Earth in relation to the Sun, but this doesn't mean there is any force on the planet capable of altering this. Similarly, there are video game bosses that, even though they move according to specific patterns, are so strong and fast that we can't effectively counter these patterns without rigorous training. However, in the case of the boss, identifying the opponent's movement patterns can still remain a key factor in guiding our training toward victory. While the example of the Sun and the Earth can be used to demonstrate that, even when a pattern doesn't open up to "flaws," identifying it can always represent an ideal way to exploit its effects to our advantage. After all, by knowing in detail how hourly or seasonal cycles work, we can also organize ourselves better to interact with them, both on a small scale (like our day-awake rhythms) and on a much larger scale, such as in planning services, transportation, and agriculture.

That being said, let's try to focus on the more practical aspect of this methodology and introduce some classic questions to guide our investigation in this direction:

- What behaviors or recurring patterns can you identify? Are there rituals, routines, or habits worth noting?
- By analyzing what is done repeatedly, can you uncover meaningful insights about what is being avoided or left undone?
- Are there periodicities or predictable patterns you could leverage to your advantage?
- Does the entity you are observing follow a routine or program that appears rigid or unchangeable?
- Do these patterns or habits reveal a lack of adaptability or flexibility? Are there signs of an inability to act or think differently?
- Could certain repetitions be deliberate distractions, designed to divert attention from something more important?
- What would happen if these repetitions or patterns suddenly ceased? Could this create opportunities or challenges?
- Are there foreseeable events that consistently lead to favorable or unfavorable conditions? Are there moments of weakness or strength you can exploit?
- Are there phases, movements, or occurrences that guarantee your safety or vulnerability? Are there specific moments when you can secure a clear advantage or must prepare to defend yourself?

Shifting focus

A shift in focus is a specific kind of "lateral approach" to our system and can be undertaken on three levels: *in relation to what we would normally do, in relation to what a certain majority would do, and in relation to what the system, or those who have designed it, "implicitly" want us to do*. While the first two represent clear invitations to seek solutions outside of our prior and social conditioning, the third offers us an additional reflection. In fact, we can say that a typical "defensive mechanism" of natural systems (and others) is to divert attention from what might undermine their integrity; much like various species that, lacking substantial strength or speed, literally "distract" potential predators with large contrasting spots on their skin. Clearly, the first thing to consider is that, while this is a possibility, it is not a universal rule. A system may, in fact, be so poorly designed that its most exploitable "weakness" is immediately obvious.

But we should also be cautious when (and it's no coincidence that we discussed this in the "Laboratory" dedicated to the "Science of Deception") the system, if endowed with sufficient complexity, can rely on this principle to "trick us." Consider, for example, a situation where what we most desire from the system is placed behind a wall that's "not too difficult to circumvent,". That obstacle may be put there with the precise intention of making us take on the challenge, scale that wall, and thereby trigger a trap. It reminds me of the well-known literary example from Orwell's book 1984, where (spoiler!) the rebellion in which the protagonists take part turns out to be part of the system itself, a device with which the Party of Big Brother can locate and neutralize potential dissidents. All this clearly invites a series of reflections on the nature, rationality, and "complexity" of decision-making and the levels of understanding within the system, for which I will, however, defer to subsequent chapters. For now, let's simply acknowledge the possibility, to be verified experimentally later, of encountering a distraction devoid of further complexities; in this case, the questions to ask might be:

- Where does everyone want me to look? Where does everyone want me to move? How can I "betray" this expectation?
- Where would someone pursuing the same goal as me look? What would anyone else here do? Can I try better? More? Can I try to do something completely different?

- Is the fact that he "holds the weapon in one hand" causing him to be too focused on that hand? Is it making him "leave the other arm exposed"?
- What are the clear limitations in the intellectual/strategic approach of these others? What do I see that they clearly don't? What can I do that they cannot or are unwilling to do?
- It's clear that many are using an imitation-based approach while I could create an approach based on the analysis of the field involved.
- What is the system "showing" me? Where does it, or whoever manages/monitors/created it, assume that I should look? Where would it like me to focus? And how can I completely overturn this perspective?
- What is the point that is most difficult to pay attention to? Which is the furthest point? Which is the most hidden, overlooked, underestimated?
- Can the system be deceptive? Alter information? Is it complex enough to create distractions, traps, diversions? Double bluffs?

Rethink the rules

This point is a "mini-variation" of the previous one. Any natural system, in an effort to "defend" itself and thus maintain its structural integrity, might (implicitly or explicitly) set rules and patterns that prevent violation, or bring it to a level that's easy to counteract. Put differently, any system that has developed at least some self-preservation mechanisms (and, we may argue, this is the case of any system that *existed for long enough*) may naturally lead you *to play by its rules*. It's not difficult, for instance, to read about authoritarian governments adopting the strategy of provoking the population into armed reactions to further justify the use of repressive measures. Or think of the serial arguers on Facebook, who trap their targets in endless discussions because they know they have the tools to rhetorically prevail over the other; these individuals likely wouldn't know what to do if the other simply began to completely avoid their rhetorical traps and ignore them.

Whoever wishes to apply this strategy must first and foremost study and understand the patterns, rules, and structures of the given context. Only then can they identify the "spaces of insertion" for their own rules, patterns, and rhythms. This strategy is employed, for example, by the pigeon from the well-known metaphor that, faced with a chess player, can only drop its droppings on the chessboard; perhaps not a great strategy for winning a match and generally one I'd advise against in human contexts. But if the sole objective were to infuriate the opponent, one could even call it a victory. The poker player with lesser game knowledge plays outside the rules but compensates with greater endurance against sleeplessness, trying to extend the game for hours. Or Alexander the Great famously played outside the rules in the anecdote of the "Gordian Knot," where he decided to cut the knot with his sword instead of untying it.

And this last episode reminds us, along with the equally legendary story of the "Egg of Columbus," that every system not only has written, evident, official rules but also rules that are considered such *because they are implicitly acknowledged as such*. However, recognizing and "officializing" these implicit rules can reveal their vulnerability. In the case of Columbus, for example, there was indeed a rule against using any tool to make the egg stand upright, and violating this would mean "losing" the game, but there were many other implicit rules that were not part of the initial agreement. Among them, the fact that the egg had to remain intact. It is precisely by leveraging this, and thus slightly cracking the base, that according to the anecdote, the Italian navigator successfully completed the challenge.

And in this case as well, let's take a look at the questions to ask ourselves to understand how to approach this:

- How many rules are truly fundamental in this "game"? Can I simply afford to break all the redundant ones?
- Can I refuse to "play" altogether?
- Is there a different way to "play" this game? Is it possible to "play" by applying completely new rules? Or at least ones that are unusual, rarely used, disorienting?
- In which rules do I have a clear advantage, or do others have a clear disadvantage? Can I somehow "shift" the game to apply those rules for everyone?
- Should I ignore certain conventions, obvious truths, common practices?
- Could I surprise through completely different *rhythms, frequencies and timings?*
- Can I use and follow some of the imposed, obvious, implicit rules to "break into" the system enough to impose my own?
- Are there constraints, rules, patterns, redundant structures that I should let go of?
- Are there constraints, rules, patterns, and structures that are less commonly used but practically more effective?

"We cannot solve problems with the same kind of thinking that created them."
(Albert Einstein)

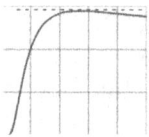

Upper limit

This principle, similar to that of "finding singularity," invites us to view our system with the awareness that none of them possesses "unlimited energy," nor can it always and in every part be at the peak of its strength, attention, and power. Instead, everything follows the inevitable cycles of "expenditure and recovery" of energy; natural rhythms of strength during which vulnerabilities may become more apparent. Lions, as dominant predators in their environment, sleep a full 15 hours a day during which they are "more or less" defenseless. Markets flourish and go through recessions, the human body experiences phases of sleep and wakefulness, and the events of our existence alternate between moments of happiness and times of loss and sadness.

However, the focus of this principle is not limited solely to the "phases" of things, but encourages us to also investigate their *structure:* it is always possible that the life, energy, and structural integrity of what interests us may falter in certain areas precisely because the resources it feeds on are not sufficient to allow uniformity in the "distribution of forces." It is the principle by which limited resources in city administrations can lead to decaying suburbs, or our heart limitations (along with the so-called mechanism of peripheral vasoconstriction) cause the tips of our toes to cool first in cold climates. And at this point, you may have already realized that this is where this principle intersects with the notion of "finding singularity": it becomes in fact now even easier to hypothesize that the system we are interested in may present "singularities" *resulting from structural or informational limitations.* This is, after all, the same principle behind exhausting a negotiating counterpart until they accept our offer or waiting for market prices to drop, perhaps due to a crisis, before buying a house.

Additionally, instead of just analyzing and waiting, one can actively create a limitation—a false attack or distraction—diverting energies to one area, then striking elsewhere. And from all these examples, it is not difficult to deduce that this applicative concept is extraordinarily valid in *conflictual or competitive* situations, where the considered system is someone-something engaged in a clear confrontation for access to spaces, energy, or resources.

Consider samurai duels, akin to chess, involving physical, biological, and mental forces like movement and concentration. Miyamoto Musashi is perhaps the most representative figure when it comes to duels with bladed weapons. He was a Japanese military man and writer who lived in the seventeenth century and is considered one of the greatest swordsmen in the history. He remains particularly famous not only for the countless literary works inspired by his exploits but also for the strategy and philosophy manual he authored, the *"Book of Five Rings."*

From what can be deduced from historical documents, it seems that during his duels, Samurai Musashi had a keen instinct for identifying *weak moments* in the opponent's rhythm, and when these weren't evident, he was skilled at wearing down and confusing his enemy until they exposed themselves. He also suggested that these openings appeared mostly in the moments *immediately before or immediately after* the "full expression of power," and by following these principles, he established the incredible record of 60 duels won, far surpassing other noted swordsmen like Itō Ittōsai, who reportedly won 33 duels "only."

And here, once again, are the questions to ask yourself to try to leverage the "upper limit" principle:

- How far can this system go, see?

- What are its limitations? Its blind spots? The resources it relies on? How limited are these resources? When do they run out?

- Does it tend to get "tired"? To overestimate itself? To exhaust itself? To bite off more than he can chew?

- Does this system expose itself, become vulnerable, open up when it is most confident in itself? Just before expressing its "maximum" power? Just after having done so?

- Is it possible, based on the previous question, to "unleash" a system's maximum power in a way that also reveals its inherent "openness"?

- Does concentrating forces, resources, and attention on a specific point or moment inherently create a "void" elsewhere or at another time?

- How can the system be confused, fatigued, or disrupted from its usual "tracks" and expectations?

- Does the system have identifiable moments of reduced alertness, diminished attention, or lower power? If so, can it be steered toward these moments?

Inevitably interconnected systems

A few pages ago, we talked about the concept of an "isolated system," adding that in most practical cases, no complex system will be truly entirely isolated nor can have meaning "on its own." On the contrary, almost everything out there will need moments and points of "opening" through which it can interact with other systems. In this context, the "scientist-hacker" might draw extremely interesting conclusions by analyzing all the points where the context of interest "opens," communicates, acquires, or exchanges information or resources. Or, where applicable, those points where it "softens," "gives trust," and must, in some way, make "necessary concessions." It can also become more interesting to analyze how it does so in cases that might be *outliers and exceptions* because, in such contexts, its priority may become opening up to "serve" the exception, maybe without much concern for maintaining a certain balance or internal integrity.

But even without necessarily referring to exceptional cases, there are many practical instances where the "need for openness" is an integral part of the system and its regular functioning. The physical guard of an opponent against whom we are fighting in a boxing match cannot remain perpetually closed, or they won't even manage to attack. Gates cannot remain permanently shut, or simply no one would be able to access the building. An email provider can't lack mechanisms for password recovery; otherwise, it would end up with more and more inactive accounts each year. Yet, it's by finding breaches in the guard that we can defeat our opponent, and often it's precisely by guessing someone's account security questions that hackers manage to gain access. In short, openness, in this case as well, whether in a *phase*, in a *part* of the system, or in *both*, can be anticipated or provoked and become a *discontinuity* to pay attention to; a revealing element of *limited resources;* an excuse to *change the rules of the game.*

To leverage the principle of interconnected systems effectively, consider these guiding questions to analyze their interactions, constraints, and opportunities:

- What other systems must this one necessarily communicate or exchange information with?

- What must it allow to pass through, enable, or overlook to function properly?

- What opportunities does it inherently create or enable? What must it provide, and under what conditions or timing?

- What elements can be offered or managed with minimal effort?

- Does this system exhibit any inherent needs, desires, or dependencies?

- What inputs or conditions must it necessarily accept or conform to?

- What exceptions or edge cases must it handle?

- Are there situations where the system's limits are tested or exceeded?

- When is compromise required for the system to continue functioning effectively?

- Under what circumstances does the system reveal vulnerabilities or weaknesses?

- What conditions might force the abandonment of core principles, structures, or commitments?

"Our strength grows out of our weaknesses."
(Ralph Waldo Emerson)

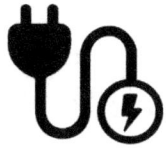

Dependency

As we just saw, in nature, most systems *can't exist in complete isolation*. It's likely that many rely on external systems for their survival, functioning, or the provision of energy, nourishment, resources, or well-being. Therefore (always raising the usual alarm signal regarding contexts where this leads to ethical and legal issues), where this "source" is more easily "attackable", controllable, manipulatable, or *where it is possible to alter and redefine the channels through which the two systems communicate*, it will be automatically possible to reach and trigger the vulnerabilities of the first system. Moreover, this argument, when reversed, leads to the fascinating principle of working on what one is dependent upon as a means to gain strength and resilience in the face of shocks and unforeseen events. This can very simply be done by minimizing external dependencies where possible and always having an alternative plan for essential resources that might fail. Clearly, managing complex systems with various micro-dependencies may require greater organizational effort, but it is possible that the cost is worth avoiding a critical risk. For example, a home alarm system that relies entirely on an external power supply, even if impenetrable, would become completely useless in the event of a simple blackout, which is why all alarm systems are equipped with an auxiliary power supply. The office handling large financial transactions cannot rely on unexpected local network failures, which is why it will be equipped with at least one alternative network to transmit data in the worst-case scenario. And this leads to the inevitable discussion of psychological dependencies, which are crucial when there's at least a partial human factor involved in the systems: because while it's true that we are made of flesh with DNA that predisposes us to fall prey to our own instincts, it is also true that, as we will see with more rigor later, in strategy rationality *should* prevail. Therefore, it is up to us to work on ourselves, reduce these dependencies and become aware of the enormous increase in potential, in terms of managing complex situations, that will result. It won't be easy in most cases, but perhaps the initial mistake is imagining that it should be.

With everything important now covered, let's dive into the usual, practical list of questions:

- What does the system depend on?
- What can it be made dependent on?
- What fuels the system, gives it power? What allows it to maintain strength and structure?
- Do the human components of the system psychologically depend on the strength, power, or effectiveness of something? Can the confidence in these elements be undermined?
- What, if removed or in turn deactivated, would render the system completely useless, helpless, defenseless, irrelevant?
- Is what the system depends on more easily attackable or controllable than the system in question?
- Are the channels through which the system communicates and exchanges resources with what it depends on more vulnerable, manipulatable, and maneuverable than the system in question?
- Once an element it depends on is removed, does the system have an alternative plan? What would it do? Would it behave in a specific way? And could its vulnerability lie precisely in this behavior?

"No matter how magnificent a building may be, if you remove the beams and columns, it will collapse."
(Gianluca Magi)

The drop hollows the stone.

"Gutta cavat lapidem," recites a Latin proverb: given sufficient time, "the drop carves even the rock." Even when it's not possible to achieve our goal through a set of impactful actions, we might reach what we're seeking through the continuous repetition of actions, possibly trivial or seemingly insignificant by themselves. Yet, each represents an "irreversible" step in the direction we desire. In this perspective, the strategist-hacker acts just like "the drop" mentioned above, constantly striving for the micro-step that will gradually "displace" the system from the "mountain" on which it is entrenched. Or perhaps it allows for the accumulation of such a quantity of strength, energy, and resources that at some point it will "inevitably counter" the system's force. Because if you can be sure that every single step carves its small part of the groove for the next day, then there will be no way to avoid facts: the groove of the previous day will add to the one of the present day, eventually allowing us to reach the other side of the rock. Granted, of course, that the system lacks the tools to "fill in" the previously carved grooves.

James Dewey Watson, the American biologist who discovered the structure of the DNA molecule along with Francis Crick, Maurice Wilkins, and Rosalind Franklin, shared an interesting principle in an interview regarding the application of "gutta cavat lapidem" to the "cancer" system and the research process to eradicate the disease. He said: *"There are numerous types of cancer. We will be able to cure some of them. And we hope to cure even more. But we must choose temporary targets. The goal is not to eradicate colon cancer tomorrow. It is to understand the disease. And there are many steps to take. No one wants to face defeat. We are happy to achieve one small goal at a time."*

From this statement, some interesting principles can be derived: first, the idea of "sharpening the tools" or the ability to "capitalize on partial progress." When it's not entirely clear how the goal will be achieved—and it rarely will be completely clear—it's important to take gradual steps that "at least bring us closer to it" as much as possible. It's possible that the final steps will emerge automatically at the end of the process, or that we won't see them at all: what matters is to try to make constant progress along the way, and to focus on *achieving milestones whose value, richness, and utility are partly independent of the chosen goal.* Or at least, try to leverage them as such in hindsight, when it's realized that the "initial plans" have partially changed. Consider, for example, what happened with Viagra, initially a medication for hypertension with "odd" side effects.

Moreover from Watson's words, and from those of anyone who has engaged with the scientific endeavor even partially, it is clear that the application of any scientific method is essentially a process of gradual "unveiling and conquest" of the unknown. Through slowly acquiring knowledge and acting accordingly, we gradually transition from being passive spectators of the forces that dominate the world, to becoming their managers and manipulators. To borrow an expression from the Chinese *"The Thirty-Six Stratagems,"* scientific progress is a gradual "exchange of roles between host and guest." Just as the ancient military manual suggested seizing control of a "home" by gradually embedding oneself in the environment until mastering its dynamics, similarly, step by step, we have moved from admiring the stars to almost reaching them, from suffering deadly plagues to extending the average human lifespan to 80 years and beyond. Considering that this path is far from complete, the thought of where we might arrive if we continue on it is breath-taking.

This entire discussion actually brings us back to the fact that the principle of "gutta cavat lapidem" can be wisely intertwined with the concept of "necessity of openness": the very structure of a system and its need to exchange resources, forces, or information with the outside can indeed lead us to actions that exploit small openings, small cracks, tiny grooves; to take steps whose repetition can lead to the creation of the "greatest advantage" we seek. Small interactions, chemical and physical reactions, necessary movements for the vital functioning of the system, can open up greater understanding of the system itself through the process of observe → study → use what has been learned to create further "openings" → observe and study more deeply.

Obviously, this approach is often used to find loopholes to exploit and drain a system: gradual uncontrolled openings, as in the case of repeated oil purchases when, due to the market turmoil caused by the COVID-19 pandemic, the price was negative, can end up depriving the original system of any sense, resource, or structure. Not surprisingly, systems susceptible to this vulnerability are monitored by engineers, mathematicians, economists, or other experts who ensure the system can't easily be "fooled" by the repetition of a partially "unsustainable long-term concession." This is typically implemented by:

- *Establishing a non-negotiable limit* on repetitions, such as stopping oil trading if its price turns negative.
- Doing everything possible to ensure that the results of the repetitions *are not cumulative.*
- Putting, as mentioned earlier, *"sand in the groove."* That would be, just as an example: on the first purchase of "negative oil" you'll get 5 dollars, on the second 2.5, on the third 1.25, etc.
- Ensuring that the positive value you may gain from a single action is merely *random* or even lower than the potential negative value.

Therefore, the effort of the "scientist-strategist," if they wish to act on such vulnerabilities, lies precisely in targeting systems where these factors have been underestimated and not considered, or alternatively; maybe they act on the factors themselves to see if it is possible to eliminate them or make them less impactful on the given context. Just like David Phillips, a civil engineer at the University of California, who realized that for every ten barcodes from "Healthy Choice Frozen Dinners" products, it was possible to earn 1,000 frequent flyer miles. Phillips discovered that the value of the miles far exceeded the value of the originally purchased products, so he bought 12,150 packages of Healthy Choice Pudding, allowing his family to fly "for free" for years.

And here too, after the anecdote, the questions:

- What "essential sub-results," if achieved, would collectively lead me to my final result?
- What "essential sub-results" could be achieved without drawing too much attention?
- Is there any "excessive" opening, any "exceptional concession" that I can iterate?
- Can an especially advantageous action be repeated? Can it be made repeatable?
- Am I capable of promising myself the willingness to commit to something even if it's slow but steady?
- What would allow me to "gradually enter" the system and alter its dynamics, step by step, to my advantage?
- What advantages, once obtained, wouldn't be reversible? How could I make them so?
- Is there an element such that, for every step I take, two steps are made in the opposite direction? If so, can I somehow influence this element?
- What can be given in exchange for something of clearly lesser value than what is bestowed?
- Is it possible that, in an exchange, someone has underestimated the value of what they are giving, or overestimated the value of what they are asking in return?
- Is it possible for environmental conditions to occur such that a particularly advantageous exchange suddenly becomes feasible?
- Has someone orchestrated a disadvantageous exchange for themselves, but with the clear intention of making subsequent ones to "recover" the loss? Can you work in such a way as to prevent these subsequent exchanges from taking place?
- Is it possible that in other places, or in the future, or under other specific conditions, what you are asking for might be given much more easily or for much less? And that what you are willing to give might take on a much greater value?

- Is there a situation where the parties involved in an exchange have completely different systems or scales of values, so that what holds particularly high value for you is of little significance to them, and vice versa?
- Is there a factor of randomness that undermines the stability and durability of my micro-steps? A factor of aging, wear, regression? Can I make each micro-step more solid and lasting?

Lack of antibodies

Although flu strains are constantly mutating, the genetic material they share can enable our bodies, upon infection, to develop at least partial immunity to some of them. A system with extensive experience handling similar challenges or inheriting "know-how" from others is likely far better prepared to counter specific strategies. On the other hand, a new, "inexperienced," or "unaccustomed" system may be more vulnerable, as it hasn't yet had the opportunity to develop the necessary "antibodies" to defend against such attacks effectively. Breaching the password system of a newly formed IT company will, in most cases, be easier than attempting to do so with a company that has been the industry leader for thirty years. (Hey, we're simply stating the facts, don't break into anyone's system, and don't make me repeat the usual ethical disclaimers every time, please!)

All of this also implies that a system which may not appear inexperienced as a whole could still have parts that are more "inexperienced," "less tested," and therefore "less suited" to defend against a particular type of approach. When attempting to "stress test" software to uncover bugs for example, it is rare to do so through the oldest features, which have likely been extensively tested. Instead, it is more likely that vulnerabilities will be found through *the more recently developed features,* which might have "lesser and poorer" defenses.

Last, it may be worth paying attention to the cases where:

- The primary defensive strategy of the system we've analyzed is to *feign a lack of antibodies, luring opponents into making reckless moves.* This tactic aims to expose and reveal their vulnerabilities. Such strategies are widely used in the animal world; consider foxes, capable of remaining motionless for hours with tongues hanging out, waiting for a predator seeking an "easy victory" to come closer and expose its neck. Similarly, disguising a state-of-the-art lock as scrap metal from the early last century could also be an intriguing "defensive" approach.

- The lack of antibodies and experience *can lead to the development of unconventional strategies,* introducing an element of unpredictability that could become a tactical advantage for the system. This is a case of what is often referred to as "beginner's luck," commonly observed in psychology-driven games like poker. A lack of experience can lead to bold, aggressive actions that disrupt the strategies of even seasoned players. These players typically anticipate a certain level of "rationality" and "caution," making them vulnerable to the unpredictability of a beginner. Alternatively, inexperience may lead to strategies based more on chance than reasoning, which, although might not lead anywhere in the long term, could, through their inscrutability and unpredictability, destabilize other participants in individual games.

Having said that, here are the questions to ask in order to "find the loophole by exploiting the lack of antibodies":

- Where is an "attack" most unexpected?
- Is there evident simplicity or tactical unpreparedness here?
- On what occasion or moment is the system least prepared for the "attack"?
- Who has less experience?

- Can this lesser experience turn into a tactical advantage for the system? Can it lead to recklessness? Unpredictability?
- Which part hasn't had the time and opportunity to evolve enough to develop defenses?
- Has something or someone here not had the chance to make mistakes and learn from them?
- Which part is newer? Which has had fewer opportunities to be tested?

Excessive force

This is the case of a very powerful military enemy that ends up neglecting their diplomatic relations and hence gets overthrown through political action. It's the *"the bigger they are, the harder they fall"* principle often illustrated in popular media. It's the event where a soccer team full of superstars ends up losing against the last team in the standings simply because they let their guard down. It's the reason why, sometimes, certain legal systems are too large and complex for those drafting new regulations to consider all of their parts and how these parts might interact. In 2005 for example, Brian Kalt, a law professor at Michigan State University, published a study suggesting a legal loophole in the portion of Yellowstone Park that belongs to Idaho, theoretically allowing crimes to go unpunished due to difficulty assembling a jury.

Obviously this theoretical principle should not encourage anyone to commit any kind of foolish act in Yellowstone Park, nor should it lead us to oversimplify the interpretation of reality through a methodology that, if applied without any critical sense, could result in the formulation of monumental absurdities (it would be entirely unscientific, for instance, to think that *certain displays of strength always mask weakness,* that the wealthy are always unhappy, and so on).

This principle should inspire us to question the initial interpretation and look deeper. It calls for a 'shift in focus' to explore whether what appears strongest or unassailable could, in fact, be the main point of weakness. This strategic approach can be summarized into these key-principles, substituting our usual list of questions:

- A powerful force could lead to an excess of organization or resources needed to keep such force strong, stable, or consistent. This can lead to lapses in attention or resources and make contradictions or conflicts more likely, like in the Death Zone.

- A powerful force might lead to an excessive focus on using such force, consequently neglecting other potentially useful factors. Here is where our ability to "shift focus" or "play outside the imposed rules" can come into play.
A practical example of this can be found in the strategies recommended for countering a particularly aggressive poker player. Such a player tends to favor bluffs, surprises, and high-impact emotional moves. Although the theoretical "psychological strength" of this approach is undeniable, it is also true that those who pursue it tend to overlook certain "reasonableness" criteria regarding what would be better to do with the cards they have in hand. Therefore, usually, having slightly higher cards is enough to "dismantle" their strategy and lead them to lose by points.

- A powerful force, as in a case similar to the previous one, could lead to overreliance on that very strength. This overconfidence might make it easier to manipulate the other into rash and reckless moves, ultimately placing them at a disadvantage.

- A powerful force could foster a dependency, whether psychological or otherwise, making the system vulnerable to destabilization simply by undermining the force itself or the relationship it relies on. An entrepreneur who has dedicated their entire life to building their business project might suddenly engage in reckless actions if they fear that this project could be threatened. This idea connects back to the first point: the inevitable cost of *maintaining what already works*. This cost becomes even more pronounced when factoring in the natural human tendency to resist change, often resulting in excessive expenditure of time, money, and resources to preserve a certain "power" status. From the perspective of a strategist employing a weakening tactic—linked to concepts like the "dopamine map" —this tendency can, in some contexts, be easily exploited to induce fatigue, scarcity, and cycles of decline, ultimately leading to a phase of weakness.

"If the opponent's numerical superiority makes the fight unequal, you must lead them to entangle themselves, rendering them powerless."
(Sun Tzu)

The human element

When discussing the human element as a potential "point of vulnerability" in a system, we can distinguish two strategically relevant cases:

- The scenario in which *only a limited set of decisions* from the human element is relevant within the "game" we have chosen to conduct. Consider all the cases where we want access to "something": a computer system, a monitored building, a social media account, specific resources. The human element vulnerability could be the person possessing the keys that ensure our access, and therefore, those "few decisions" that lead them to provide us with these keys might be enough, after which their role can be considered concluded.

- The case in which the human element *a fundamental decision-maker* within the game we've chosen to conduct within the system itself. Consider a card game, chess, or a complex system of diplomatic relations. In this scenario, the entire game, or at least a significant part of it, unfolds on the level of human decisions and how they influence both other decision-makers and external elements.

While the second case will be addressed in a dedicated chapter, this section focuses solely on the first set of circumstances. In practice, many "hybrid" cases may fall between the two, making dynamics useful in one context often applicable to the other. For simplicity, however, we'll use this classification as a kind of "compass."

One important note: the issue of the "human element" is perhaps the most delicate so far, and considering human beings as mere tools for executing our "scientific hacks" rarely leads to positive outcomes. However, if you are still in doubt, I invite you to reread what was said just a few pages ago about "not making a mess"; a small admonition to which I wish to add the idea, in which I strongly believe, that understanding the nature of certain procedures can always be useful for developing a certain level of *self-defense* against them. On the other hand, I firmly believe that outright censorship is archaic, ineffective, and counterproductive. Not only does it hinder our ability to understand how malicious hackers, unscrupulous marketers, and fraudsters might exploit vulnerabilities, but it also reflects a flawed reliance on *"security through obscurity"*—the mistaken belief that hiding information provides protection. In today's world, such an approach is untenable; even if we chose to avoid discussing these topics, countless other websites and social media platforms would still disseminate them.

Returning to our main point, this strategy aims to influence the human element, guiding it toward a controlled set of decisions that pave the way to the desired outcome. This approach is particularly effective when such interventions are more practical than directly addressing the system's complexities or defenses. It is especially relevant in highly complex or well-defended systems—at least relative to our practical capabilities—where key individuals can effortlessly bypass these complexities and defenses.

This is what is commonly referred to as *"social engineering,"* an approach made popular by former hacker and computer expert

Kevin Mitnick. A particularly famous case can be found in the story of Stanley Rifkin, which I often love to cite. It was the fall of 1978 when Rifkin was working for a company tasked with setting up a backup system for banking data. When he entered the control room of Security Pacific National Bank, where all the terminals were located, he took note of the procedures—ostensibly so that his backup system would synchronize with the normal one—and meanwhile took the opportunity to glance at and memorize the slips of paper on which employees wrote the security codes for bank transfers. A few minutes later, he headed straight for the phone booth, inserted a coin, and dialed the room's number. He then impersonated Mike Hansen, an employee of the bank's foreign desk. Using the name Mike Hansen and leveraging the security code he had just obtained, he instructed the transfer of $10,200,000 through the Irving Trust Company of New York to the Wozchod Handels Bank in Zurich, where he had already opened an account. A few days later, Rifkin flew to Switzerland, withdrew the money, and handed over $8 million to a Russian agency in exchange for a bag of diamonds. He then returned home, passing through customs with the stones hidden in a money belt. He had pulled off the biggest bank heist in history, all with the simple use of two tools to exploit the human element: a phone and a deceptive confidence.

Today, due to increased awareness, it's nearly impossible to execute similar bank schemes. However, situations where key figures compromise complex systems still deserve attention. It's a bit like that scene in the animated show "The Simpsons," where the owner of the nuclear power plant, Mr. Burns, tries to enter his maximum-security bunker through various ultra-complex security systems, only to end up in a small room with a broken wooden door leading outside, through which a stray dog has just entered. Same level of vulnerability, with the difference that one is a cartoon gag, while the other is a principle considered acceptable and "normal". Consider for example all those entrepreneurs who believe that investing in training programs against social engineering for their employees is ultimately of "secondary importance": the risk of one of them giving away crucial keys for the entire business, thereby compromising it entirely, is "scientifically" just around the corner.

Strategy Lab - Scientific Self-Defense

This very small, final "strategy lab" is nothing more than an excuse to highlight something absolutely fundamental from a practical standpoint: everything you've read so far regarding the natural vulnerabilities of natural (and other) systems can be reinterpreted through the lens of "And now, how could an external element attack one of my vulnerabilities?" Indeed, this chapter can become an incredibly valuable guide on how to scientifically enhance our resilience against all those "attacks" that, in the form of problems, unexpected events, or even direct assaults, could inevitably compromise our existence (and, I would feel inclined to add, this should probably be your sole approach to such principles).

By reducing our dependencies, as previously mentioned, we become more resilient to unforeseen events. By not giving too much, we can prevent external elements from taking advantage of us. By keeping a keen focus on the finiteness and "cycles of strength" of our resources, we can avoid finding ourselves overwhelmed with too many problems at once. And so on. It all comes down to having the "scientific humility" to "surrender" to reality and thus recognize our gaps, our recurring "openings," our weaknesses; consequently, in a mirror-like manner to what was said in the paragraph about "excess of strength," we can automatically transform these potentially debilitating factors into the "seeds" from which to cultivate our greatest sources of success and happiness.

Stories of "Scientific Champions"

Galileo Galilei: the first "science hacker"?

Born in 1564 in Pisa, Italy, Galileo was a pioneer in many fields of science, from astronomy to physics. However, it was his challenge to the geocentric model of the solar system, the accepted idea that the Earth was at the center of the universe, that put him on a collision course with the ecclesiastical authorities.

"Armed" with a telescope, a relatively new instrument at the time, Galileo made a series of revolutionary discoveries. He saw the mountains on the Moon, observed the moons of Jupiter, and noted the phases of Venus; all observations incompatible with the geocentric model.

Nonetheless, presenting such evidence was no walk in the park for Galileo: claiming that the Earth orbited around the Sun, rather than being the center of the cosmos, was considered heretical; a direct challenge to the authority of the Church and its sacred texts.

But Galileo did not stop. Despite warnings and threats, he continued to share his discoveries, driven by an unwavering belief in empirical science and the need to spread the truth, regardless of personal consequences.

This rebellious spirit cost him dearly. In 1633, he was tried by the Roman Inquisition and declared "vehemently suspect of heresy." He was forced to recant his beliefs and spent the rest of his life under house arrest.

However, his indomitable spirit and thirst for truth left an indelible mark on science. Thanks to his tenacity, solid foundations were laid for the heliocentric model to become the accepted norm, but above all, his methodology established systematic foundations that would guide future generations of researchers, literally changing the human world forever.

Galileo could perhaps be called (or, in any case, I'll do it as I love the idea) the first "science hacker," a scientist who defied the rules of his time and dared to question the "vulnerabilities" of the established order. This story is a powerful testimony to the struggle for truth in an era of obscurantism and to the importance of using all means available to fight against corrupt, malevolent systems eager to impose truths authoritatively. Because systems eventually collapse, especially if they're built on fragile and ideological foundations; reality, on the other hand, ultimately "prevails by its own strength" over everything else.

V - Towards the Core of Truth

"We know so little, yet it's incredible to note HOW MUCH we know, and it's even more incredible to think about how so little knowledge can give us so much power."
(Bertrand Russell)

In this chapter, we will delve a bit more into the practical theme of *discovery*, the unveiling of hidden or less obvious truths. The scientific investigation techniques we will explore together will in fact help us understand how to "scientifically" approach the problem of answering questions such as: *"How can I attempt to unveil a truth I am unaware of, or at least arrive at one that is probabilistically closest to it?"* This, hopefully, may prove to be incredibly useful in any practical daily situation, from analyzing which company might have the greatest potential for you to accept a job offer, to reconstructing your family history with some scattered old photographs, or even attempting the "far from scientific" task of figuring out which gift your partner would appreciate the most for their birthday. But let's begin right away:

Start with some assumptions

A fundamental principle for working towards reconstructing a "truth" is, albeit simple yet often challenging, to accept starting from *"I don't know."* How often, recently, have you heard someone make such a statement? I'd wager that the answer would suggest not too frequently, given that admitting one's ignorance seems outdated, a sort of cultural taboo. This has naturally been reinforced in a social media era where one feels compelled to have an opinion on every subject; and often, we even feel obliged to express it with absolute conviction and certainty, as if trapped in communication styles that penalize any form of doubt or uncertainty.

Nevertheless, doubt and uncertainty, as anyone who has reached this point should now well understand, are essential factors in any advancement toward truth. This brings us back to the fact that, indeed, in any investigative process, we must be ready to replace these initial *"I don't know"* with sets of *hypotheses:* responses that are partially incomplete, uncertain or undefined. But above all, we must understand that, depending on the contexts and the tools at our disposal, even our "final" truths might remain "cursed" by at least a partial element of uncertainty. However, this needn't concern us too much, as our discussions on importance of *pragmatism* should have made it clear by now: we don't always need perfect or complete truths to take a "calculated risk" and act.

But how should these hypotheses be gathered or constructed, exactly? Everything, and by now this shouldn't surprise you anymore, starts from an observation of reality that is as *objective, verifiable, precise, and extensive as possible* (more data, more sources, more reliable information) and is adjusted based on an awareness of our errors and biases: primarily, the "confirmation and ostrich" biases that are always lurking, making us collect only the data that conform to our initial expectations. All of this is aimed at arriving at a collection of reasonably verifiable notions and data that come both from our observations and from "quality sources". Then, this might be enough for us to start formulating valid hypotheses or might require a subsequent stage: that is, working on what has been gathered by processing ideas born from investigating, delving deeper, recombining, and extracting relationships from what has been collected; this can be done with techniques such as the "dopaminergic map," or through models like those we will see in a few pages.

Finally, once we have gathered several hypotheses, it will be essential—due to inevitable considerations of "economy"—to prioritize those that are probabilistically most valid, as well as the ones offering the greatest utility and reasonableness from a cost/benefit perspective.

For example, imagine you're trying to organize family photos chronologically. You may need to determine which of two images, depicting unknown ancestors, is older. A preliminary hypothesis might be to assume that the more damaged photo is older—plausible, but not certain—further supported by the observation that the clothing in the photo appears to belong to an earlier era. This assumption might suffice as a starting point for your research. Later, however, you could gather additional information by: 1) Asking your grandmother questions, and 2) Investigating the evolution of photographic techniques. From this, new insights might emerge:

- According to your grandmother, the most damaged photo is "probably" more recent because it shows an uncle who was not yet born when the other person was alive; however, she is not entirely sure about it.
- Those seemingly more "remote" clothes are simply because they belonged to ancestors who lived in the countryside.
- The most damaged photo appears to have been printed using a more advanced color technique.

Given these final notions, the idea that "more ruined equals older" could be simply dismissed (which may require the "courage" to set aside something that initially seemed good or valid to us), and the new probabilistically more valid hypothesis becomes the opposite. This, clearly, could be refined by further research.

We will explore later in this chapter how to perform some of these "hypothesis refinement steps" in more detail. For now, however, I hope to have clarified, or reiterated, some of the fundamental premises underlying any investigation: starting with the awareness of not knowing, proceeding with the understanding of the necessity to work with partial truths. Then analyzing reality, working with probability, refining hypotheses with new data or results from possible experiments, and stopping when it becomes "pragmatically reasonable" in light of our goal. This alone offers an incredibly valuable foundation of guidelines to address a wide range of practical problems, even if they remain somewhat vague or incomplete.

Wild pruning

Before exploring further methods for processing hypotheses, let's take a moment to talk of *filtering and selecting them*. Given our inherent limitations in resources and processing capabilities, it is clear that any workflow will benefit from *narrowing* the boundaries of the context being analyzed as much as possible.

A key principle to consider here is that, before even "selecting" what is probabilistically more valid, useful, and reasonable from a cost/benefit perspective, the process should begin by *prioritizing the removal of everything that objectively fails to meet these criteria*. Eliminating irrelevant information early will in fact enormously simplify the identification of significant details. Which recommends, in any investigative process, to first exclude all nonsense, every definitely wrong answer, any foreign element, all irrelevant data, any illogical presence; to "eliminate" everything that can quickly be verified as such, but above all, to abandon any *unnecessarily complex hypothesis*.

This principle, which you may have already heard of, is called "Occam's Razor" or the "principle of economy." It is considered one of the foundations of modern scientific thought and can be summed up as *"it would be foolish to work with more elements than necessary."* To explain it even better: it's the reason why, when we flip a switch to turn on a light bulb, we simply define the phenomenon as being caused by electricity flowing through the wires, without necessarily attributing it to the work of some invisible elf turning on the light, all while a second elf creates the illusion in our minds of a rational mechanism based on physical principles.

However, to move towards hypothetical fields that are a bit less absurd, this is also the spirit that drives historical reconstruction: although we can never be completely certain, given multiple accounts from historians describing, for instance, the unfolding of the Punic Wars between 264 and 146 BC, it is more "economical" to think that these wars actually occurred, rather than to hypothesize that all those historians were simultaneously part of a conspiracy. Of course, if a number of historically verifiable documents should "suddenly" validate the conspiracy theory, we could revise our assumptions and continue the investigation; the point, as always, is to see if this "emergence of disruptive novelty" comes from a rational comparison of verifiable sources, or from an attempt to satisfy our own "bias," desperately seeking a conspiracy, an architect, a hidden truth.

This method, in essence, can not only streamline our initial assumptions from the outset, making the overall process much simpler, but it might even be enough to automatically help us reach our "final" truth. For instance, imagine having a multiple-choice quiz with four options: if you're fortunate enough to dismiss at least two of them as absurd, further reflection on the remaining two might even lead you to certainty about which one is correct.

It is possible to observe that this tool is repeatedly adopted by the character Sherlock Holmes in the stories of Arthur Conan Doyle, who, as mentioned earlier, was one of the first to apply a pseudo-scientific method to criminal investigations in literature: by starting from his encyclopedic knowledge and observation, Holmes was able to define a set of hypothetical solutions. Then, through the evaluation of facts and deductions about them, he gradually performed a "wild pruning" until reaching the famous "brilliant creative abduction" that provided a possible explanation of how the crime unfolded. Alternatively, if narrowing the context revealed illogical solutions, it indicated additional possibilities to consider.

Because even this latter option deserves consideration when "pruning" our range of possibilities. If what remains after this process still proves useless, irrelevant, or illogical, it's a signal to *step back and reassess*. We might have "pruned too much," or perhaps we began with an unrealistic or irrelevant set of options. Holmes' famous adage, *"Once you eliminate the impossible, whatever remains, no matter how improbable, must be the truth,"* comes in fact to mind. While somewhat simplistic given the complexity of certain systems (or the endless plausible hypotheses for some phenomena) it remains a valuable principle for countless practical situations, especially those with well-defined constraints.

Interactions and receivers

The method based on the study of "interactions" is another way to formulate hypotheses, and can be summarized as: when the nature, constitution, or laws that animate something are uncertain or difficult to study, one can try to understand more by analyzing *how this element interacts with other entities, things, or people*. It's not coincidental that a few pages ago we talked about how, in reality, almost no system makes sense "in itself." On the contrary, many complex systems have, as their *raison d'être*, the purpose of opening up and exchanging information and resources with the outside world. Hence:

- **We can attempt to derive hypotheses from the nature and outcomes of these interactions.** When one element triggers a change in another, it is possible that some part of the nature, structure, or characteristics of the source is instilled in the "receivers" of its influence. This transformation, when these changes can be traced and investigated, turns the receivers into custodians of valuable information. This is how the ancient Greek mathematician Eratosthenes was able to measure the Earth's circumference, for example: a simple calculation of how the Earth's spherical shape was affecting the angle of the Sun made it possible to achieve an extraordinarily accurate astronomical measurement, over 2000 years ago and without any advanced technological instruments. It's also how experienced poker players try to uncover their opponents' bluffs: the supposed emotional pressure from bluffing can sometimes lead to clear and recurring micro-expressions, and identifying these can unveil an entire game strategy.

Clearly, a whole cosmos of "scientific" problems can arise when we try to evaluate the precise set of laws through which phenomenon "A" (unknown) should influence phenomenon "B" (which we observe): if the law meant to link the two is misunderstood, unreliable, or variable over time, we may reach incorrect conclusions. Think of the bluff example: if our opponent actually showed a "nervous tic" the first time they bluffed, but that tic also appears when they have an excellent hand, then unreliable laws could lead to "terribly wrong" behaviors and courses of action. However, given what we learned about *probabilities, biases in hypothesis formation,* and the *role of experimentation,* the possibility of error in the hypothesis formation process should no longer intimidate us.

- **We can try to introduce new "receivers" into the context**, which interact with the source system and thereby reveal useful information. A spy dispatched among "enemy lines," a probe sent into space, or a drug that triggers specific chemical responses within a biological body are just a few examples of strategies for information gathering through the "introduction of receivers." This approach, in contexts where the receiver is *reliable* (consider a space probe that breaks shortly after launch or a sensor that records a false positive) and the practice itself is *not so disruptive as to alter the behavior of the system itself* (think of when an enemy spy is discovered and allowed to continue operating, but only receives false information) could be one of the best ways to gather hypotheses when direct evidence is lacking.

Similarity and Past

Another method for formulating hypotheses about reality involves analyzing *the behavior of similar elements (or the same element) in past situations.* Imagine, for example, having to take an exam at university without any idea about the behavior of the professor who will evaluate you: you could simply investigate by looking into how that professor behaved while examining other students. Consider how economists and mathematicians attempt to predict future trends of variables like the number of infections during an epidemic or a country's GDP variation. Experts create graphs or models of these variables or similar ones in comparable contexts. They then identify patterns or laws to help predict future trends. However, this continues only until singularities, exceptions, or other factors intervene, partially invalidating the previous analysis.

From all this, it is indeed easy to understand that this methodology has three weaknesses, of which every mathematician, scientist, or economist is well aware:

Number one, *in the case of analyzing "similar" elements or events, similarities do not imply identity.* An epidemic, for example, may have a completely different course from a past epidemic. Thus, while it's acceptable to consider similar phenomena for study purposes, one must also always consider whether the differences introduce variables significant enough to impact the ongoing analysis substantially.

Number two, *the reliability of a law depends on the stability,* and consequently the predictability, *of the system from which the law was derived.* The more we examine very complex and unstable real systems, the more likely it is that even the most detailed models may prove to be "quite unreliable" in providing us with reliable hypotheses.

Number three, there is always a serious risk of *overfitting,* where models become excessively tailored to past data, capturing not only the underlying patterns but also the anomalies and random fluctuations. Such models, while seemingly accurate in hindsight, may perform poorly when applied to new, unseen situations. For instance, an economic model meticulously adjusted to fit historical data might fail to predict future market behaviors, especially when unprecedented events occur.

Even though these weaknesses may evoke a Lovecraftian "horror of the void" in some of you, let's not forget the positive side. By focusing on the essential patterns without getting lost in the minutiae—by remembering that *while history rhymes, it doesn't always repeat*—we can create models that offer meaningful insights. If we remain mindful of not over-interpreting past data (hence avoiding the trap of *overfitting* where we mistake noise for signal) we can steer clear of common pitfalls. Given, for example, a system similar to the one being considered, or if conditions remain stable enough, we could formulate more probable hypotheses that guide us reasonably well.

The alternative to this uncertainty, after all, would be the ability to always predict *every future event* in detail; and although the hypothesis that this might one day be possible with the arrival of sufficiently complex technology is intriguing, the truth is that in the current state of affairs, *science seems to cautiously suggest that achieving perfect predictions in every field will never be possible*. Therefore, there are certain aspects of life where we must accept uncertainty, hoping for favorable outcomes. This, considering how much our human nature also craves mystery and adventure, is a far more comforting truth than it may initially seem.

Build the hypothesis model

In the previous section, we discussed the possibility, based on observations of reality, of constructing a predictive model—essentially a representation of that reality that defines its behavior. This, in turn, allows us both to make potential predictions about it and to formulate hypothetical plans on how best to act within it.

Let's take a closer look at a possible scientific procedure for "model construction" of this kind, which is highly useful when the unknown we are seeking is related to a phenomenon and its potential developments. Of course, this book is not the place to explain how to build detailed mathematical models for scientists or academics; nor can we hope to provide in a few lines the tools to make elaborate predictions about very complex systems. Therefore, let's consider what follows as a "playful prototype," to be refined each time according to the specific needs of the case considered:

Observation and data collection: If you find yourself in a studyable field, try first to observe, as suggested earlier, the evolution of elements, structures, behaviors similar to those of your interest. Or look at past patterns. Pay attention to the conditions in which these patterns and behaviors occurred. Is there a regularity, frequency, or rhythm? A recurrence of patterns? A clear appeal to rules or occurrences? Do quantities vary following a specific pattern? Are there specific conditions A that generate an effect B? Are there clear differences or discontinuities between similar events? And what causes the latter?

But above all, are there conditions that science or human knowledge provide us with, where we objectively already know there's a higher probability of generating certain values, certain occurrences? Correlations with pre-existing mathematical, physical, chemical, or medical laws? Human tendencies to behave one way or another? Intrinsic characteristics of what we are observing?

As seen a few pages ago, it will be enough to answer these questions by building and collecting a dataset and information that is extensive, based on rigorous formalizations where necessary, and extracted from reasonably reliable and verifiable sources (or perhaps even partially validated through an initial set of experiments). In this way, we will already have an excellent set of "building blocks" with which to construct our model.

Schematic representation: Let's now try to create a small table where we can record our data and information; this typically becomes useful when documenting the occurrence of events or variations in quantities, especially in relation to time, if relevant, and/or under given conditions.

For example, if you want to understand how to reduce your need for extra snacks during the day, it might be helpful to track your daily calorie intake for a few days:

- Based on your mood
- Based on the amount of physical activity done
- Based on other "potentially" influential factors such as hours of sleep, the presence of migraines, and similar issues.

Additionally, I'll throw out a suggestion: at this stage, you might also benefit from converting your table into a graph, or using some software that extracts useful occurrences and patterns for you. However, this field is so vast that it can't be easily condensed into just a few lines. In general, all spreadsheet applications like Excel, Numbers, and similar provide the ability to perform such operations with great ease. To delve deeper into the topic, search engines can always be your "best friends."

Drafting the first hypothetical model: At this point, simply try to observe your tables, your charts, or the way you've represented this information. Depending on the phenomenon, you might realize that one or more quantities tend to vary in a way that can be mathematically described through a formula (for example: given your current lifestyle, you tend to gain two kilos per year). Or you might notice the emergence of patterns, rhythms, recurrences, or possible causes (while always remembering to pay attention to the bias of correlation vs. causation). Perhaps, once again referring to the previous example analysis, comparing how many calories you consume in relation to the quality of your days might reveal a pattern suggesting that it's precisely on days when you sleep less than seven hours that you end up binge-eating afternoon snacks, in an attempt to "mitigate" stress and tiredness resulting from poor sleep habits. Or when you eat too much, do you tend to sleep worse?

However, *be cautious not to force patterns where they may not exist*. Overcomplicating your model by trying to fit every minor fluctuation in your data can lead to misleading conclusions. This risks falling into the same trap we touched on earlier when we talked about *overfitting*—an overzealous attempt to extract meaning from every fluctuation, blurring the line between true patterns and mere noise.

Clearly, on the other hand, your tables or graphs might not reveal any obvious patterns, in which case further observations, measurements, data collection, and systematizations may be necessary. Additionally, further hypotheses about "possible" conditions not yet considered might be required. For instance, if our extra snacking on Wednesdays has no link with mood, sleep, or physical activity, could it be related to our stress levels? Or to the quality of our relationships? Ultimately, it will be up to us to determine the value of continuing the research process, especially considering the expenses it entails, and whether it might be better to stop and "yield" to the temporary inability to extract a useful law, pattern, or "regularity."

Change and retry

With the "change and retry" technique, we delve into the part of the chapter more closely dedicated to *experimental* techniques. This method can be particularly useful when we already have *multiple hypotheses about the causes of something* and, therefore, want to understand which one is most likely to be valid. For example, we might not know which component in a machine is causing it to malfunction. Alternatively, we might be unsure which neighbor is making a dreadful noise at four in the morning, waking us up in the middle of the night. We have an idea (hypothesis), but we want to verify which one is correct.

This method involves trying to remove or replace (or "wait" for it to be removed or replaced) a potential cause and see if the effects we're interested in somehow disappear or change. If they do, we've *possibly* found a potential cause and thus have something to work with. For example, when a connection between your TV and your new PlayStation isn't working, the first thing a technician might try is to test the same TV with a different console and the same console with a different TV. If this doesn't yield useful information, they would examine the physical connections between the two components with different TVs and consoles.

When the presence (or absence) of one element in the combination reveals it as the "sole and sufficient" cause for the overall system not working, it becomes easier to understand where an intervention might be wise.

Interestingly, the "change and try again" technique is one of the "worst enemies" of the *"confusing correlation with causation"* bias. When you have an event A that might cause B, and you replace A with C, the "magic" of the method reveals itself: if B continues to occur largely unchanged, it's likely that A and B were merely correlated—perhaps due to a shared cause or mutual influence—rather than A being the direct cause of B. However, if replacing A with C stops or significantly alters B, you may have found strong evidence to confirm your causation hypothesis.

Note all those "mostly" and "could"; they're obviously resulting from the fact that reality is in most cases more complex than it might initially appear. This method, as described, is in fact just a mere simplification, with its main weakness arising from the assumptions that:

- *At least one* of the elements you've identified as a potential cause of the phenomenon *is actually responsible.*

- We have the concrete opportunity and enough time *to make or see a sufficient number of "exchanges."*

- There aren't any further complexities, such as *multiple contributing causes* or cases where the behavior of the considered elements *varies significantly or unpredictably over time.*

So what should you do in the presence of one or more of these circumstances? The answer to this question actually offers suggestions applicable to all existing experimental techniques: first, *don't rely solely on a single method of investigation and experimentation*, but try to integrate it with all other cognitive tools at your disposal. Second: if possible, repeat the application of your method (or methods) of investigation, because if you notice patterns emerging over multiple applications, then even a partial model of reality might start to emerge. And thirdly and lastly: when repeating your "experiments," apply your methods starting from initial conditions that are as reproducible, verifiable, and similar (if not identical) to each other as possible. This is what is defined as "recreating laboratory conditions," a premise without which it would not be possible to obtain reliable results from any experiment.

Obviously, recreating similar conditions won't always be feasible in the case of complex real-world phenomena, especially when we have no real control regarding their initial state. However, even without reproducible and verifiable laboratory conditions, we *could* still gather more reliable data if:

- The variations do not significantly impact the investigation's outcome.

- We understand how these variations affect the results, and hence we can recalibrate or adjust our conclusions accordingly.

Ripples in the water

When a system is unable to reveal useful information on its own, it may be necessary to introduce an external element or impart a new signal that works as a detector. This serves to differentiate it from merely being a receiver (as discussed a few pages ago) and represents something that "provokes" or "stresses" the system in question. By altering its structure, operations or patterns, it forces the system to "reveal" its behavior.

This practice is often used in conflictual "games" where the revelation or secrecy of information is crucial, and therefore each contender has every interest in "leaking as little as possible" about their intentions. In these cases, numerous military strategy manuals, like the well-known *"The Art of War"* by Sun-Tzu, suggest using a decoy or a scare tactic so that the enemy is led to partially expose themselves, thus revealing some of their behavior. Alternatively, in a different context, if a software is "broken" and we don't have direct access to its code, we can "provoke" it. We can stress the software itself to cause the error to manifest in various ways until it becomes clear which component or context is creating the problem.

Clearly, once again, much here depends on whether the provocation reveals useful truths rather than *excessively distorting* the information. However, if it is established that we are within these boundaries, or at least capable of assessing the extent of this distortion, then a skillful game of "stones thrown into stagnant water" at the right times and rhythms can prove to be an extraordinarily powerful weapon in our "arsenal" as explorers of the unknown.

Waiting for Spring

As Warren Buffett said, you can't just *"have a baby in one month by getting nine women pregnant"*: a slightly extreme phrase, but its absurdity serves as a reminder: sometimes, all we need is patience to respect a system's natural balance and allow it to reveal the information we seek.

This can typically happen for two reasons. First, considering the natural cycles of systems and their shifts between "high" and "low" energy phases, what we seek may naturally enter moments of reduced alertness, caution, or defense, revealing what was previously hidden. Consider what happens in historical research: today, it is possible to reconstruct in detail events from just 30 or 40 years ago because we can work with documents such as minutes of private conversations and confidential meetings that were completely inaccessible until very recently, due to obvious security or privacy reasons. Barring any sensational information leaks, it is likely that in 30-40 years, we will say the same about today's events, gaining a completely new perspective that, unless we find ourselves personally in the "corridors of power," would be simply unimaginable today.

But the strategic wait can also be due to the fact that the foundations and premises necessary to reach the knowledge we need have not yet been created, or it is not yet convenient to do so. Imagine trying to sequence the human genome during the Roman Empire: even if they could understand the concept (and they likely couldn't), they only had tools for crushing herbs and making ointments.

This, of course, is another "extreme" example, and no one expects you to wait for millennia to figure out something like the secret ingredient in the main sandwich at the corner deli (which I generally wouldn't advise you to investigate). However, it serves as a small demonstration that time alone can transform something perhaps initially unimaginable into something later evident; thus, it is theoretically possible that the set of information you need will eventually become apparent due to the spontaneous evolution of things. This doesn't even necessarily imply a completely passive wait in the face of our unknown. Often, the best revealing games can in fact be orchestrated through the right balance between action and waiting: think of waiting after having "disturbed the water" or "introduced a new receiver." The key is to ensure that every action remains respectful of the evolutionary timing of certain systems, or their parts. Consider, for example, the case of a person who needs time and comfort to reveal information within a psychotherapeutic process; in these cases, a few well-considered "ripples in the water," like well-aimed questions, combined with deep respect for the natural reaction to these "disturbing" elements, could indeed lead to the flow of information the psychotherapist requires.

Strategy Lab - Scientific Intuition

Let's take a moment to explore one of the least scientific ways of understanding reality: intuition. As noted early in this book, intuition can be seen as a major "source of nonsense." Prone to every imaginable bias, it once led us to believe deities controlled lightning or that the Earth was the center of the universe. Reality, more often than not, is counterintuitive.

But how could we still make it an "ally"? First, let's take into consideration that the beauty of intuition lies in the fact that it is the most economical hypothesis processor at our disposal, with all the obvious considerations that follow. Foremost among them is the much-emphasized necessity to always refine one's intuitive hypotheses in light of what reality and experimentation suggest. Interestingly, recent neuroscientific studies have shown that intuition is not just a cerebral phenomenon but involves the entire body. Our "gut feelings," for instance, are rooted in the enteric nervous system—a network of neurons in our gastrointestinal tract that communicates with the brain. This suggests that our body can reveal insights that our conscious mind may not immediately grasp.

This also implies the impossibility of deriving anything good from one's intuition when dealing with concepts completely detached from any verifiable or falsifiable reality, such as metaphysical ones: since there is no possibility to confirm or refute these intuitions, they can be "always true" or "always false" depending on what we decide, and thus we are light years away from a rational process of analyzing reality.

Moreover, one should not underestimate the fact that, given its immense cost-effectiveness, there could be countless life situations where intuition proves to be the only hypothesis generator capable of providing us with something meaningful. Consider, for instance, contexts where time is a critical factor and we must make a good decision with few tools or very limited information at our disposal. In high-stress situations, our body's rapid responses—mediated by both the brain and peripheral nervous system—can alert us to dangers or opportunities before we consciously recognize them.

But since also these responses are often imprecise and we would probably never purchase a car simply because it's very cheap, let's take a moment to understand how to improve our intuition quality. In this regard, a fundamental distinction can be made between the intuition of those who are at least partly aware of their biases and that of those who are not at all. Although both are the result of what Daniel Kahneman called "System 1 thinking"—fast, automatic, and often unconscious mental processes—it is possible that a certain level of awareness of how some biases influence this process can "realign" its outcome towards greater accuracy. Cognitive psychologists in fact seems to suggest that metacognition, or thinking about one's own thinking, allows individuals to recognize and mitigate the impact of biases on their intuitive judgments.

Consider, for example, trying to intuit whether you can "trust" the abilities and intellectual honesty of a newly acquainted colleague at work. Our brain may have unconsciously absorbed information about a series of subtle nuances in the colleague's behavior, tone of voice, and the content of their phrases; all things that, once processed, could already produce a clear "yes/no" result regarding their reliability. Moreover, physiological responses—like a tightening in the stomach or a sense of ease—might be signaling your body's appraisal of the situation.

And here's where the essential difference between a "biased" intuition and a "calibrated" one comes into play: the former might have given excessive weight to certain natural preconceptions, like the fact that this person resembles a lazy colleague or that their tone of voice reminds us too much of our infamous alcoholic uncle. However, if we are already aware of our tendency towards these biases, we can recognize them in advance and attempt to "subtract" these variables from our considerations. By doing so, we aim to focus on more grounded and rational elements we have observed; perhaps we noticed them picking up a wallet from the floor and placing it on the desk of the colleague who lost it, which would theoretically indicate someone more honest, more reliable. In this way, we align our intuition with empirical evidence, creating in our minds something much closer to the "brilliant creative deduction of Holmes" that Eco spoke about, rather than the frenzy of a witch doctor.

From all this, one can also easily conclude that intuition can neither be, nor ever will be, a complete substitute for solid knowledge, observation, and rational analysis of the reality around us. However, it can be, depending on the context, an excellent complement. This is not only from a purely pragmatic standpoint—considering that we might not always have enough time for more in-depth studies on what interests us—but also in terms of mutual reinforcement. An in-depth knowledge in a particular field can indeed add more "significant" building blocks to the processes through which our intuitions are formed. For instance, an experienced doctor can more easily "intuit" a diagnosis for a patient even before the necessary clinical tests are performed, partly because their extensive knowledge and prior experiences inform their gut feelings.

From which comes the final, straightforward yet valuable advice: train your intuition. While imperfect, it remains a powerful tool when combined with critical thinking and rigorous analysis. This training involves not just accumulating knowledge but also developing an awareness of one's own cognitive processes and bodily signals. By paying attention to both mental and physiological cues, we can enhance our ability to make swift yet accurate judgments. However, our intuition should always operate within the framework of evidence, ensuring it enhances understanding rather than misguiding it.

Stories of "Scientific Champions"

Barbara McClintock and the Importance of the "Quest for Truth"

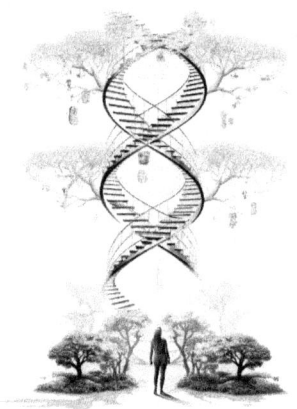

Barbara McClintock was born in 1902 in Hartford, Connecticut, USA. She was a solitary and introverted woman, yet incredibly passionate about science. She chose the field of genetics, which was still partly unexplored at the time, and became one of the few women to attend Cornell University in the 1920s.

While many of her colleagues followed well-established research paths, McClintock ventured into uncharted territories. Studying maize between the 1940s and 1950s, she observed that some DNA fragments moved from one part of the genome to another, influencing the plant's development. These "jumping" DNA elements were a revolutionary discovery; until then, it was believed that genes were fixed and immutable in their positions along the chromosome.

Initially, some of her colleagues dismissed her discoveries as mistaken or irrelevant. But McClintock was undeterred. With patience and determination, she continued to gather data, driven by her deep belief in the truth of her observations and the importance of seeking a truth, even when that truth challenged conventional thinking.

After decades of isolation and rigorous work, the scientific community finally recognized the importance of her discoveries. In 1983, McClintock was awarded the Nobel Prize in Physiology or Medicine for the discovery of "jumping genes," officially known as transposable elements.

Although initially opposed, her discovery not only rewrote the rules of genetics but also opened new frontiers in the understanding of the human genome and genetic diseases. Today, transposable elements are recognized as a crucial component of DNA, making up a significant part of the genome of many species, including humans. They are involved in a variety of cellular functions and can influence the evolution of the genome.

McClintock's work is a clear example of the importance of persistently pursuing truth through scientific research, even when such research contradicts accepted knowledge. It is the story of a scientist who, through her dedication and brilliant insight, changed our way of looking at biology on a molecular scale, pushing science in a completely new direction.

This story is indeed an extraordinary testament to the power of perseverance, and the love for science and the pursuit of truth. In a world where new ideas can be easily dismissed or ridiculed even by those who claim to be "champions" of the same science, her journey shows us the enormous difference between what is truly science and what is merely "the opinion of some scientists."

VI - The keys to Mastering the Future

We should know it by now: as good scientists, we must accept the possibility that analyses may fail, unknowns may remain unknown, and we may have to surrender to the *unknowability and unpredictability* of things. Consider, for the most basic example of all, the game of casino roulette: no matter how much we analyze the physical structure of the roulette, gather data, find patterns, consult with those who are "experienced," or even use our intuition, we will always and inevitably conclude that no strategy can be profitable in the long term. However, our scientific advantage lies precisely in getting a *more precise* understanding of the uncertainties and unpredictabilities of the universe. By getting in fact *better measurements* (see previous chapter) on the "no's" that reality shouts, we can sensibly and rationally adapt our operational strategies accordingly. But let's see how.

How to predict the future

Short answer: as mentioned a few pages ago, *you can't*. A slightly more complete answer: you can't, but let's see what mathematics can offer us to better understand and manage the degree of uncertainty of future events. Let's hence take a closer look at what *probability theory* tells us about it. If these are topics you are "more or less" already familiar with (I have also discussed them in another one of my books, the "Speed Math Bible"), feel free to skip this section and move on.

For those who have decided to stay: probability theory is *the branch of mathematics that studies the likelihood of events with uncertain outcomes*. Given that one defining characteristic of our species since the dawn of time is an obsession with the thrill of gambling, it is not surprising that the earliest known probability studies involve dice games. The first significant work on probability that we know of is in fact *"Liber de ludo aleae"* by Gerolamo Cardano, written in 1526 and published over a hundred years later, in 1663.

The very concept of probability, on the other hand, while well-known in common language, does not have a single definition in mathematics; however, for the sake of simplicity, we will try to quickly define it as *the measure of how likely an event is to occur*. And above all, given our deeply pragmatic nature, let's immediately try to answer the question: "How do I calculate it?"

First of all, as mentioned when we discussed the cost/benefit ratio, probability can be expressed in values between 0 and 1, a convention typically used in mathematics, or in percentage points from 0 to 100, which is probably more intuitive given its frequent use in everyday language. These are clearly the same range of values, and these two are just different ways to represent them. Therefore, for example, one can say that a certain event, such as the idea that the precious set of grandma's china would break if you threw a medicine ball at it, has a probability of 1 or 100%. Similarly, an impossible event, such as the idea that you could win the lottery without buying a ticket, has a probability of 0 or 0%. These, along with all the values in between, define the probability of occurrence of any real or simulated phenomenon.

Now, given for simplicity that we will use values ranging *from 0 to 1* for our calculations (although you can easily obtain the corresponding percentages by multiplying them by 100), we can say we can measure the probability of any event as follows:

Through a mathematical evaluation: this type of assessment can be done in an *"a priori* way," which means through equations based on the structural analysis of the phenomenon.

This type of analysis is more immediately applicable to somewhat less complex phenomena and often begins with the principle that, given an event that can occur in "n" possible ways, each with the same probability of occurring, the probability for each specific event to happen is 1/n (one *out of* n, or one *divided by* n).

When for example we talk about rolling a standard die, it can obviously result in 6 possible outcomes, each with an equal probability of occurring. Therefore, each face has a probability of 1 divided by 6 = 0.1666..., or approximately 0.17, of appearing. Alternatively, if we multiply it by one hundred, it's 16.667%, or roughly 17%.

In the case of flipping a coin instead, the process is just the same: given the two possible outcomes, each has a probability of 1/2 = 0.5 = 50%.

Through a statistical evaluation: this type of assessment is carried out "retrospectively," requiring the study of similar past events and a reliable collection of data on the matter. By adopting such an evaluation, we can say, simplifying it enough, that the probability of an event occurring *is equal to the number of times it has occurred, divided by the total number of times it could have occurred.* For example, one might state that the probability of an airplane having an accident is equal to **"Number of air incidents" / "Total number of flights"** within a well-defined time frame.

Even now, it is possible to note that this method, although more immediately applicable to complex phenomena than the previous one, has the disadvantage of being impractical if there isn't a sufficient amount of significant data available. Additionally, it risks leading us into endless debates about what is meant by "significant data"; for example, does it really make sense to consider flight data from a hundred years ago? And what about data from a completely different model? We saw in the previous chapter, when we talked about building predictive or descriptive models, that complexities are many.

Anyway, even here, without any claim of providing rigorous definitions that would require separate books, much can be gleaned from what has been said in the previous chapters: it's better to work with *as much data as possible*, rely on the use of *credible sources* and differentiate *signal from noise*. For everything else, it is probably best to leave the details of specific cases to industry professionals.

In theory there's much more to be said on this topic, particularly concerning advanced statistical methods and the philosophical interpretations of probability; most notably, *Bayes' Theorem*. This theorem provides in fact a framework for updating our beliefs based on new evidence, blending prior knowledge with observed data. However, delving into the intricacies of Bayesian reasoning is beyond the scope of this book. For those interested in exploring this fascinating area further, good ol' Google and Wikipedia can be our best friends here.

Let's hence return to two little things worth analyzing within this context, always with the aim of better understanding some slightly more complex phenomena. First, the probability of something **not happening.** In other words, very simply, in order to obtain the probability that an event *will not* occur, you just need to calculate *1 minus its probability of occurrence*. For example, what is the probability of rolling 6 with a die? We saw it: about 0.17. Then, the probability of *not* rolling a 6, meaning rolling any of the other numbers? **1 - 0.17 = 0.83**, or about 83%.

Then it can be useful to analyze cases where we want to understand:

The probability of multiple events happening together, which is simply *the product* of the probabilities of the individual events.

For example, if I want to roll a die and flip a coin and my goal is to determine the probability of the event *"Getting heads AND the die shows a 6,"* then the probability will be **1/2 x 1/6 = 1/12**, which is 0.08 = 8% probability.

The probability that at least one of several events occurs, instead is simply 1 minus the probability that *none of them* occurs. Intuitively after all, "at least one happens" is the same as saying "it's not true that none happen."

So first, for each of them, you take the "probability of non-occurrence," of each event which, as seen earlier, equals *1 minus the probability of occurrence*. Then, to determine the probability that all these "non-occurrences" happen simultaneously, you simply multiply them together. Finally, you subtract the result of this multiplication from 1. But yeah, since this process can sound a bit complex, let's better understand it through an example:

- Let's consider the case mentioned earlier, where we want to roll a die and flip a coin, but this time we want to determine the probability of the event: *"It's heads OR the die shows a 6."*

- **Probability of getting heads:** 0.5. Probability of NOT getting heads: 1 - 0.5 = still 0.5

- **Probability of getting a 6:** approximately 0.17. Probability of NOT getting a 6, as seen earlier: 1 - 0.17 = 0.83

- **Probability that neither of them occurs:** 0.5 x 0.83 = 0.415

- **Final step:** we need to subtract the amount we just obtained from 1, giving us our final result: 1 - 0.415 = 0.585, or 58.5%. In other words, slightly more than 50%, which was probably predictable intuitively, given that 0.5 is the probability related to the coin event alone, and adding the condition of "getting a 6 on the die" will simply increase the event probability by a bit.

The Formula for Success

Assuming that we should at least intuitively understand the relationship between the numerical value of a probability and its corresponding "intuitive" degree of uncertainty (0 being something impossible, 1 being certain, 0.01 being quite unlikely, and so on), let's now try to see how what we've discussed so far can represent one of the most powerful mathematical tools for making "successful" decisions.

First, let's try to create the "mathematical proof" of the fact that *resisting and persevering* can represent the best possible strategy of action in nearly any situation with uncertain outcomes.

Let's imagine we're in a situation where we repeatedly "try" to achieve something, until we achieve at least *one* success that satisfies us. It's like when we ask different people to "go out" until we find the love of our life; or when we have to go through job interviews until we're offered the position we aspire to; or even when we try to pitch to potential clients until we find someone willing to pay well for our service, and so on.

In uncertain situations, the probability of failure in a single attempt is often much higher than the probability of success. Despite this, how do we still manage to succeed? For some, the answer is intuitive, but let's try to prove it mathematically.

The first consideration we can make here actually draws from the formula for "calculating the probability that at least one of several events occurs."

Indeed, even if our probability of success on a single attempt is extremely low, let's say 1% or 0.01, the more we repeat these attempts, the more we make it highly unlikely that at least one of them won't succeed. For instance, if we manage to repeat our highly improbable (1%) attempt 100 times, our new probability of achieving at least one success (and no, it's not 100%) will be 63.4%. If the attempts increase to 200, the probability will rise to about 87%.

The mathematical steps to achieve these results derive precisely from what we saw a few pages ago on calculating the probability that "at least one of several events occurs," and we leave their development to the reader's curiosity. What we really want to focus on is how repetition has transformed that "measly" 1% into an almost certainty of success. Which, if you think about it, is really "not bad"; it's almost like extracting a precious stone from an almost-depleted mine.

However, it's enough to look at the reality of things, and at the moment when excessive insistence with a crush has led to being blocked on every social media, to understand that *perseverance doesn't always pay off*. This leads us to conclude that this mathematical model alone can't describe a successful strategy for every real case, so we need to make additional considerations.

Let's say we call the representative of a major publishing house to ask if they want to publish one of our books, with our probability of success always being 1%, and their initial response is a very impolite: "I have no intention of publishing your garbage." It's very likely that on the second call, the publisher would be even more annoyed than before, so our probability would drop to 0. And no matter how many times we try, probability will stay 0. Therefore, trying to apply this value to the previous formula, we'll find that if the probability of success for individual events remains zero, the probability of success in the long term will also remain zero. Which, simply put, means that *there's no number of attempts that can turn that "no" into a "yes."*

The first crucial aspect in our "game of attempts" hence is to ensure that the probability of each attempt remains constant or even increases with each try. How to achieve this will largely depend on the situation encountered, but a good general guideline could be to *focus on improving one's technique with every attempt* and to *explore environments with high potential and internal variability*. This would not only allow us to maintain high chances of success in individual efforts but also mitigate the risk of a failure in one attempt lowering the probability of success in the next. For instance, in the recent case, one could consider calling another department of the same publisher, or even try with a different publisher; additionally, it might be "mathematically" beneficial to see if we can enhance our persuasion technique, better present our idea, or introduce any other "enhancing" factors (such as learning a lesson after each attempt on how to improve the next one).

In short, we have expanded our "formula for success" with: increase your attempts, improve or learn something more with each try, and direct them each time towards situations that are sufficiently (if not more) "rich," variable, and full of potential. Given enough freedom of action, it will be really difficult not to "hit the mark" at least once.

But in reality, we can't even stop here. Let's say we are door-to-door vacuum cleaner salespeople. The formula suggests that if we have a good product, learn from rejections, and try selling to enough people, we will likely get at least one *yes*.

But... what if this means we still end up with an unacceptable result? For example, finding ourselves selling only one vacuum cleaner every six months, and perhaps even spending a lot of money on gas with each attempt?

This section, after all, could hardly be called a "formula for success" if we didn't also consider such cases; indeed, at the beginning, if you remember well, we made a significant premise, namely that *the expenses of the various attempts were "sustainable"* and especially that even a single "victory" would be enough to consider any expense worthwhile. But what if reality were different from this?

The answer to all this lies in further expanding the mathematical model of our situation, this time with the so-called "expected value formula," which is a value we can define as the average expected gain on a single attempt:

Expected Value (EV) =
Probability of gain per single attempt x gain amount - Probability of loss per single attempt x loss amount

This formula, in essence, highlights that it's not just important for the probability of success to be close to 1 in the long term (which under certain conditions will be "almost always true"), but it's also crucial that over the long period, the potential average loss is negligible compared to the potential average gain. So, to clarify a bit further, if we wanted to input hypothetical values related to the street vendor's case into this formula, we would have that:

- Probability of profit on a single attempt: that is, the probability of selling a vacuum cleaner according to our statistical data; let's assume so far we have observed 1 success out of 20 attempts and that this value, for simplicity, is also predictive = 1/20 = 0.05, or 5%
- Profit from the sale of a vacuum cleaner: €200
- Probability of loss: certain, therefore 1. Every trip involves a non-negotiable expense.
- Average cost of a trip: €5

And in this case, we would have:

$$EV = 0.05 \times 200 - (5) = 10 - 5 = 5$$

In other words, after subtracting expenses, each trip to a potential client on average earns us 5 euros. Which, yes, means that repeated attempts will "inevitably lead to at least one success," but it's a value that can also immediately guide us toward analyzing more complex issues: does this average earning, for example, justify all the effort? Could we reduce expenses to increase this average? Increase the margins per individual sale? Increase the number of trips we make in a day? Improve our success rate by learning how to be "more persuasive"? Change work entirely?

But to better understand this principle, let's apply it to a situation where we might find ourselves at the Monte Carlo Casino and want to decide whether to bet on red or black at the roulette. If we're dealing with a roulette wheel that has red and black numbers plus zero and double zero, the probability of winning by betting on a color is about 0.48, or 48% (half the numbers, minus zero and double zero), while our probability of losing would be 0.52, or 52% (the other half, plus zero and double zero). And if we decide to repeatedly play by betting five euros each time, by plugging these values into our formula:

EV = 0.48 (probability of gain) x 5 (gain amount) - 0.52 (probability of loss) x 5 (loss amount) = 2.4 - 2.6 = -0.2

The result is a negative number: in the long run, there is no way to win the game, but on average, we will lose 20 cents for every 5-euro bet on red or black. Much worse than with the vacuum cleaner, and therefore, as per the "grandmother's recommendation," it's a game in which it is not worth placing bets at all.

Sure, once again we find ourselves facing the usual problem: everything is beautiful and simple when it comes to dice and roulette, but measuring these factors in detail in everyday situations can be incredibly difficult. However, even here, it goes without saying, something can be done by taking inspiration from what's mentioned in the chapter "no numbers, no party": try to use the most "quality" tools to carry out your measurements and, where that is not possible or convenient, *try to rely on estimates*. For instance, if you don't have precise quantitative criteria to evaluate your gain or loss, simply try to rate them based on how much each event would satisfy or harm you. The same goes for probabilities: if you cannot calculate them, can you at least sense them? For example, is an event completely uncertain? Estimate its probability at 0.5. Is it quite unlikely? Set it to 0.2. Is it practically but not entirely certain? Set it to 0.85. Or, try asking yourself: how many times, more or less, might this thing happen out of a hundred times?

This formula can not only help us determine in advance whether a game is worth playing, but it can also provide us with some crucial "scientific" advice to consider in our personal journeys.

First: this formula is somewhat the mathematical demonstration that *if you have nothing to lose, it's always worth trying*. In life as in science, mathematics will always tend to reward initiative in conditions of negligible risk. Even when the probability of success is extremely low, adopting the mindset of "persevere and try again" is simply the best path to pursue an advantage despite uncertainty. Of course, there will always be "invisible" and often immeasurable costs, the first being the time invested in our attempts. However, assuming for simplicity that these costs are also "reasonably sustainable," we can say we have an "universal principle of common sense" to cherish.

Let's further expand on the topic, reconnecting with the series of "mathematical principles for success" we discussed a moment ago, and let's try to complete the list. Up to this point, we have tried to:

- *Increase the number of attempts*
- *Improve with every attempt*
- *Always guide ourselves towards situations that are reasonably "rich," and intrinsically variable/diverse.*

Given our new goal based on making the "expected value" as high as possible through our strategies, we can also append the principle: "and try to minimize the cost of each attempt as much as possible." Or, to be even more precise, "make it negligible compared to the potential gain." This could also imply striving to aim high and ensuring, wherever possible, that the successful attempt is worth the entire effort made up to that point. The more you can make the cost, effort, and time and resource consumption of each individual attempt irrelevant in comparison to the gain, the more you will find yourself in the situation described at the beginning of the chapter, where perseverance will inevitably be your most valuable "scientific ally."

"Nothing in life is to be feared, it is only to be understood. Now is the time to understand more, so that we may fear less."
(Marie Curie)

We could end this subchapter with a lovely, "scientific" image that I like to use whenever I want to reinforce the idea of persevering on one's goals. Imagine an open box, divided into 100 sections and containing 50 balls, each neatly placed in every other section; now imagine closing the box and shaking it repeatedly until the balls are chaotically distributed within the container: some sections will contain three, others four, some none at all. Now, think of each section as an attempt and each ball as a success; you can even reduce them to 20 or 10 if you wish, but the point is: the distribution between successes and failures along your path will inevitably be chaotic, because chaotic are the mathematical laws governing complex situations. There could be very long sequences of completely empty sections, but if you continue to explore, you might suddenly encounter areas with 4 or 5 balls all together. Getting "lost" in the empty sequences as if they were the whole story is, quite simply, *the worst mistake we could ever make.*

Strategic and tactical variability

Let's take another step back from the previous chapter and remove the assumption that we have multiple attempts at our disposal. In short, we find ourselves with a certain degree of uncertainty and various possible scenarios, and we want to understand what science suggests to make the most out of not necessarily the long-term, but *first and foremost the single event.*

A fundamental principle that can be derived from the chapter where we discussed flexibility and "necessary variety" is that we will tend to handle uncertainty better whenever we equip ourselves with some "internal variety": with strength, structure, and strategic-tactical capacity capable of "reasonably managing" a variety of situations *equal to or greater* than those the system might present.

But let's better explain the concept with an example: let's say we have to face a university history exam where we might be asked questions about three possible topics: the Late Middle Ages, the Early Middle Ages, and the Renaissance. These three historical periods represent, barring surprises, the greatest variety the system can offer us. Therefore, if we study enough to achieve an "internal variety" such that we are able to answer any possible question about these historical periods, we will almost certainly be able to manage the uncertain situation. If we study even more, preparing, for example, thematic connections with other historical periods, we might even achieve an unexpected excellence that earns us top honors. If, on the other hand, we do "less" and prepare only for the Early and Late Middle Ages, nothing prevents us from still managing the situation in two out of three cases, although it does not "mathematically" guarantee our success. This, mind you, unless we already possess a level of knowledge of historical processes that allows us to excellently improvise possible talks on the Renaissance; or perhaps to sneak a peek at the book during those couple of minutes the professor, without assistants, needs for a "quick break" to the bathroom. Improbable, but certainly not unrealistic.

The difference between the "internal variability" achieved through prior study and that achieved through the ability to improvise in the moment actually leads us to the formulation of two possible strategies to enhance our ability to manage uncertain situations:

1. Strategic Variability: This is everything that results from the *preemptive analysis* of possible scenarios and the *systematic preparation* of strategies to separately address each one of them, or at least the most likely and risky ones. This will be a worthwhile goal, especially if our situation and its potential developments are fairly predictable in advance, if there is sufficient lead time to prepare, and if the risk of not considering possible scenarios is significant (especially relative to the cost). The latter factor can be simply calculated through the expected value mentioned earlier: by calculating the *probability of a negative scenario occurring,* multiplied by *the damage it would cause,* and then comparing the result with *the cost needed for preventive measures.*

Example: Suppose there is a 5% chance that the ceiling of our bedroom will collapse because of construction with poor quality materials. If it collapses, the damage to objects might amount to ten thousand euros. Damage to people, however, could be incalculable. Repair costs: 5,000 euros. So, on one side we have 5,000, and on the other a probability of 0.05 multiplied by an "incalculable" damage value, which we could approximate to infinity. Solution: fix that stuff and don't take it too lightly. Find a way to repair it at lower costs, perhaps, but don't risk losing things you cannot afford to lose.

2. Tactical Variability: That is, everything that allows us to bring together tools, behaviors, and structures to face uncertainty in the moment, regardless of actual events. This is unsurprisingly a good objective, especially when the situation cannot be easily studied in advance; therefore, the most sensible thing to do is to "be in the moment" and try to extract from it everything useful that our intuition and observation can suggest.

Clearly, except for straightforward situations or those with absolutely predictable uncertainty, such as a coin toss, the extreme complexity of situations we will face implies that an ideal approach includes *strategic factors for everything that is "economically" feasible to plan in advance*, as well as *tactical factors that "cover" everything else*. This might involve preparing for an exam by studying the answers to the most likely questions and learning how to "improvise," if possible, a response for everything else. Or, to apply it to a more general context, approaching life by "planning just enough" and allowing everything else to be "effectively managed" by our intuition, adaptability, and resilience.

Playing with fire

In this chapter, we have so far examined countless real-life cases where a situation can have different possible outcomes. However, in this section, we will introduce a new variable: *the presence of a decision-maker*. This could refer to any complex entity—be it a human, machine, animal, or system—that makes choices, whether those choices are rational or irrational, deliberate or instinctive. And this scenario, as many will know, falls within the set of cases analyzed by the mathematical theory developed by researchers like Von Neumann and Morgenstern, known as *game theory*.

Given that game theory is an incredibly vast and complex branch of mathematics, we will simplify its discussion "just enough" while attempting to understand, hopefully, all the most critical and relevant cases.

Human and Elven games

A common example often used to introduce game theory is the "prisoner's dilemma," a classic case of a "2 x 2 game," where there are two contenders, each with two possible strategies. This is theoretically not too limiting, both because the simplicity of these games is ideal for illustrating the rules that also govern more complex cases, and because a vast number of practical problems can be solved through the analysis of 2 x 2 games. Negotiations between parties, disputes, legal cases, exams, tennis or chess games, university tests, taking a penalty in soccer, and even simple "me vs. potential crossroads" situations, are all scenarios where one must opt for one of two possible major groups of choices, due to an element that will do the same.

But let's return to the case of our prisoners: there are two men, A and B, who have been investigated for a significant crime and taken into custody. Neither has the opportunity to understand what the other will do. However, they have TWO possible choices: deny or confess, and depending on what they do, the police will act in a specific manner. Namely:

- If both deny having committed the crime, *each will be sentenced to only one year in prison.*
- If one confesses and the other denies, *the one who denies gets ten years in prison, while the one who confesses will be free.*
- If both confess, *they will each be given five years in prison.*

Games of this type are typically represented using a table called a "payoff matrix." In this matrix, the columns correspond to *the decisions of one decision-maker* (e.g., Player A), while the rows correspond to *the decisions of the other decision-maker* (e.g., Player B). Each cell at the intersection of a row and column represents *the outcome* when both players choose those specific actions, showing the benefits or losses for each player. Therefore, in this case, for example, the cell at the intersection of "B confesses" and "A denies" will depict the numerical model of the situation where "B goes free and A serves ten years in prison." Thus, our entire situation can be represented in the table as follows (note that the prison years are shown as negative values, as they are a drawback):

	"B" confesses	"B" denies
"A" confesses	A = -5. B = -5.	A = 0. B = -10.
"A" denies	A = -10. B = 0.	A = -1. B = -1.

Let's now imagine being "A" ourselves, with our goal being to minimize our number of years in prison without caring too much about "B" (thus, for simplicity, as is said in these cases, assuming the game is "non-cooperative"), and we find ourselves wanting to understand what is the best thing to do, without knowing the intentions of "B" at all. Therefore, we could represent the same table, but only with "A's" indices:

	"B" confesses	"B" denies
"A" confesses	-5	0
"A" denies	-10	-1

The sensible first step here might be to verify if there exists a so-called *"pure dominant strategy,"* a strategy that can always guarantee us the best possible advantage, regardless of the other person's choice. If you think about it, this could easily extend to numerous everyday situations. How many times, for instance, do we have no idea what will happen, and we end up asking ourselves, "Okay, but what move can get us a benefit, or at least prevent us from harm, *no matter* how things turn out?" That's how we could call that in game theory context: we look for *pure dominant strategies*.

And this approach, in a case represented in a tabular manner like the one just seen, can be realized by trying to understand *if each of our values in one row is always greater than the value in the other row.* Therefore, here we consistently have that -5 > -10 and 0 > -1; in other words, in the case where both of the others' choices are sensible and possible, it will always be better for us to *confess*.

But let's remove the possibility of adopting pure strategies by changing a condition: suddenly we find ourselves in the land of the elves and the agreement has been altered according to elven law. The new conditions are: if both confess, or if both deny, a pat on the back and no jail time for anyone because that's how it's done among elves who appreciate harmony between the two decision-makers. However, if you confess and the other denies, you'll still get two and a half years in prison because elves hate lies. According to elven jurisprudence, the ten years remain if you deny and the other confesses. Let's therefore modify our table in response to the new rules learned:

	"B" confesses	"B" denies
"A" confesses	0	-2.5
"A" denies	-10	0

In this case, we have that 0 is indeed more than -10, but also that -2.5 is less than 0, so there are no pure dominant strategies. However, let's see what we can do to try to maximize our advantage.

Indeed, in these cases, assuming we have absolutely no idea what the other party will do, one possible solution is to adopt a "mixed" strategy. This involves a partially random strategy which, as such, does not mathematically guarantee the maximum advantage in a single "game," but can offer some probabilistic assurance of success over multiple attempts. Additionally, this approach benefits from possessing "substantial intrinsic unpredictability," another factor that, as we will soon see, can represent a clear advantage in any competition between decision-makers.

And so, the mathematical tool capable of guiding us towards the application of this methodology is the calculation of the so-called *value difference* for each row (which is essentially a form of "expected value"). That is (and pay attention, as mathematical steps are going to be slightly more complex here):

- Take *the largest number in each row* and *subtract the smallest number in the same row.*
- Place this value *on the other row.* This will represent the value difference for that line.
- Now convert the value difference into a percentage of the total to get the *probabilistic value difference.* To do this: First, *add together* the two value differences you calculated earlier. Then, *divide* each value difference by this total. Finally, *multiply each result by 100* to express it as a percentage. For example, if the value differences from the previous step are 10 for Choice X and 20 for Choice Y:
  - **Their total** is 10 + 20 = 30.
  - **For Choice X:** 10/30=0.33. Multiplying by 100 gives 33.3%.
  - **For Choice Y:** 10/60=0.66. Multiplying by 100 gives 66.6%.

These values provide us with three possible indications on how to act: **the first, simple one**, based on adopting the choice with the highest index.

Second option: in the case of repeated games, make a choice based on a frequency relative to its probabilistic value difference. In the example just illustrated (with the two arbitrary values 10 and 20; we will return to the "elven jurisprudence" case shortly) this would mean choosing option X, valued at 10, one out of three times (33.3%, or 0.33 of the cases), and option Y, valued at 20, two out of three times (66.6% or 0.66 of the cases). However, by adhering to a predetermined rhythm, this might expose us to the disadvantage of extreme predictability.

Third possibility: whether or not these are repetitions, you can use a *random* tool. It could be a coin, a die, or even the second hand on a watch or a cell phone clock (since looking at it at any random moment is like drawing a random number from 60). Calculate the chances percentage on *value differences* and try to recreate an event that has the same probability of occurring as one of the two values in question. If that event occurs, make the decision associated with that value; otherwise, choose the other option. Which is actually much easier done than said: for instance, again, in the previous case, we have choices with values of 33.3% and 66.6%, meaning roughly one-third and two-thirds. This means you could check your phone at any given moment and make the first decision if you're within the first 20 seconds (the first third) of the current minute, and the second decision if you're in the remaining 40 seconds (one of the other two-thirds).

Alternatively, if your probabilistic *value differences* were 50 and 50 (0.5/0.5), you can flip a coin and simply move based on whether it lands on heads or tails; or alternatively, you can look at the clock and decide based on whether you're in the first 30 seconds of the minute or the last 30 seconds.

But maybe you have two probabilistic values like 10% and 90%. The decision, once again, can be made using a clock: choose the first option if you're within the first 6 seconds, and the second option if you're in the remaining portion of the minute.

Let's now try to understand everything even better by returning to the example of Elvish jurisprudence from earlier.

For the choice of confessing: take 0, the highest value, subtract -2.5 and you will get 2.5. Apply this 2.5 to the other row, which is the choice of denying.

Now, for the option associated with our denial: take the usual 0, subtract -10 from it, and you will get 10. Do the same, and this will be the value difference for the choice of confessing.

Already from now, we can see that the choice to confess has a higher index, which suggests it might be the option with the highest probability of maximizing our gain. But if we wanted to go even further into detail, we could evolve it into the "probabilistic value difference"; and thus, we would realize that we end up with a "pie" of 12.5, where one part takes 10 slices and another part takes 2.5. From this, the part that has 10 owns 80%, and the other, by subtraction, 20%. In short, it might be advantageous for us to confess in front of the elves 80 times out of 100 (or, simplifying, 4 out of 5); which can equate to confessing if the hand is in the first 12 seconds of the minute (12 is 20% of 60). That, modeled in a table, would be:

	"B" confesses	"B" denies	PVD
"A" confesses	0	-2.5	10 (80%)
"A" denies	-10	0	2.5 (20%)

The Arrival of the Forest Spies

In the previous case, we started from an assumption, which fortunately cannot always be taken for granted, that we have no information whatsoever to predict the other's behavior. Let's assume, then, that this assumption falls apart and we receive a tip-off from our elven informant. We might find out that the other person is much more inclined to deny rather than confess. Consequently, the game could change significantly, and as a result, the optimal strategy to adopt could also change considerably. Let's see how.

A first method that could be used is to consider the opponent's most likely choice as good, eliminate all other columns from the table, and make our decision accordingly. In this case, we have:

	"B" denies
"A" confesses	-2.5
"A" denies	0

And so here is a table from which it won't be too difficult to understand that the best thing to do will be to deny in turn.

But suppose our informant is not 100% sure about what they've told us; or for some reason, given the risks, we don't feel comfortable ruling out the other strategy. In that case, we can act in a slightly different way.

As you may have already guessed, in this case, we will apply the concept of *probability*. More precisely, we can multiply the values in each column by their likelihood of occurrence and act accordingly. Let's assume, for instance, that the other person will almost certainly end up denying it. We decide to estimate that "almost certainly" with a probability of 0.9, meaning 90% or nine cases out of ten, and we adjust the table as follows:

	"B" confesses (10%)	"B" denies (90%)
"A" confesses	0	90% of -2.5 = -2.25
"A" denies	10% of -10 = -1	0

Now that we have recalculated our values based on a proper assessment of uncertainties, we can simply behave as before: if there is a *pure dominant* strategy, let's adopt it. Otherwise, we'll apply a *mixed strategy*.

In this case, given the absence of "dominant" rows, we always end up with a mixed strategy. Therefore, we calculate the value difference with our new numbers and find that, for the first row, 0 - (-2.25) = 2.25, which goes on the deny row. For the second row, 0 - (-1) = 1, which goes on the confess row.

	"B" confesses	"B" denies	PVD
"A" confesses	0	-2.25	1 (≈30%)
"A" denies	-10	0	2.25 (≈70%)

Once again, we find ourselves in a situation where, regardless of the method used, merely having information about the other party has completely changed the game. Previously, we found that in 80% of cases it would have been better to confess; now, denying will give us an advantage in more than double the number of cases.

And considering how little the method changes if the players or their possible choices are also more (just add more rows or more columns; or, in the case of more players, have more tables), it can also be easily deduced that, whatever the type of game considered:

- Obtaining information about the possible behaviors of others *can completely change the significance and potential success of our choices*. Therefore, even just knowing something about the other person's character, attitude, or tendencies might allow us to create strategic "tables" that are closer to reality, enabling us to make much better decisions. For example, in the most basic scenario, could you assign a higher probability coefficient to the idea that the other person will choose the option that is most advantageous for them, if such an option exists?

- Having a strategy that includes at least *one totally unpredictable or random choice* forces the opponent to "examine a smaller table." Unpredictability or randomness, if properly thought as a part of a strategic plan, can lead to an advantage. Whether the opponent actually has a table in front of them or not, the unpredictable option nullifies their ability to perform preliminary calculations to counter that specific move, thus forcing them into an "improvised tactical response".

- Having a "tabular" and therefore detailed view of the ongoing game provides us with a valuable guide to understand, if we can, whether it is worth *trying to influence others' behaviors* and strategies to our advantage. The very fact that we are working with an instrument prepared in a sufficiently rigorous manner can provide us with equally reliable indications on how to direct this persuasion attempt (assuming there is a practical possibility). In the very first case of elven jurisprudence, for example, when we had no idea what the other would do, it would have been in our interest to *lead them to confess,* given how much greater mutual advantages would have been compared to the other possibility. But we could also consider leading them to deny if we had already decided to deny ourselves and wanted the certainty of zero years in elf prison.

These principles apply symmetrically and, when combined with the concept of "variability," they explain the dynamics of any competitive or conflictual situation. For instance, introducing a degree of "information distortion" about ourselves, or influencing how others perceive that information can shift dynamics in our favor. Consider chess: while you can't hide your moves, you can psychologically disrupt an opponent, reducing their ability to interpret the game effectively. Similarly, any action whose purpose is to *diminish the influence others may have over us* can further skew outcomes in our favor.

A deeper understanding of the other party is often *key to maximizing outcomes*. This principle holds true on both large scales, such as the games driving market trends, and small ones, like deciding who will handle grocery shopping during a conversation with a partner.

Understanding games, in short, means gaining *greater control over them*. Not an "absolute" control, of course, as many will still depend on other factors such as luck, skill, and power dynamics; however, as mentioned a few pages ago, the more the disparities between the resources and strengths of the contenders are reduced, the more a superior, broader, deeper, and diversified strategic insight can literally make all the difference.

Strategy Lab - The Science of Winning

What we will do in this very last "strategy lab" is to gather "fundamental questions" that encapsulate all the observations made so far, to guide us in all competitive or conflict situations we may encounter in our lives.

The idea, as usual, is: first, if the list of questions doesn't interest you at the moment, skip this section and move on. If that's not the case, review them and try to understand which best apply to your situation. It could be an athletic competition, an arm-wrestling match, a negotiation with an elf judge, or an attempt to persuade your partner to spend more time with you over the weekend. Since there is an infinite number of contexts that can be modeled mathematically through "games," there truly are no limits to how many practical situations can benefit from this exercise. The important thing, of course, is to always play with fairness and common sense.

Understand the game and the "numbers" involved in the stakes

- What exactly are you aiming for? What is your opponent aiming for instead?
- Are your interests conflicting? Is it possible to conduct your activities separately so they aren't?
- Would it even be possible to cooperate in *sharing* what's at stake?
- Is it mostly a game that requires particularly aggressive strategies, where you must prevent the other from gaining any relevant advantage as much as possible?
- Does the aggressiveness of the strategy require being particularly fast? Unpredictable? Bringing together resources that the other person doesn't have?
- Or perhaps the game requires you to be slow, cautious, and gain a small "piece" of advantage at a time?
- Is it necessary to be defensive? Offensive? Aim for both? Secure a "safe haven" from which to defend and then counterattack?
- Can you outline the current situation through a table or a scheme?
- Could you understand the best course of action right now, regardless of what others might do? Is there a "pure," dominant strategy worth adopting here?

"Scientifically" analyze the other

- Can you somehow predict what the other person will do? Do you have any idea how they have behaved in similar situations in the past? Do you know how similar people behave?
- Does the other person follow habits? Do they tend to follow predictable patterns?
- Does the other person have a readable character? Predominant instincts? Myths? Rituals it cannot escape? Strong desires pushing it in a direction? Tendency to fall prey to biases? Tendency to prioritize certain decisions over others?
- Is the other person seeking power? Money? Security? Social validation? Consistency with past statements?
- Is the other person uninformed in a specific field? Do they have doubts about a particular topic?
- Does the other person tend to believe in specific authorities, figures, or symbols? To follow them?

Try to enter a state of "heightened variability"

- How does the other person move? How far do they usually go?
- What would you never do? What would they never do? How is this a problem?
- Where is the other person completely unable to look? Where is the other person not accustomed to looking?
- What would you never imagine?
- What are the other person's limitations in terms of energy, vision, time?
- What constraints is the other person inevitably bound by? Physical? Moral? Spatial? Temporal? Role-related?
- Do you know things that the other person doesn't? Are you sure about these things? And are you sure the other person is unaware of them?

- Does the other person know things you don't?
- Is the information you receive about the other person reliable? Does it come from credible sources? Could this information be manipulated? Is it biased?
- Are you ready for the worst-case scenario? Considering its likelihood and the cost to "protect yourself," is it worth it for you to be prepared?
- Could the other person learn things about you that would put you at a disadvantage? Are you also a victim of biases, habits, or patterns that make you predictable? Is this a problem?
- Is the other person clever? A deceiver? Could they lead you to believe they'll decide one thing just to steer you in a particular direction? Or are they honest? Transparent? Predictable?
- Has the other person already shown that he does not prioritize honesty and transparency on other occasions or with other interlocutors?
- What interests are surely much more significant and valuable to the other person than honest and transparent communication? What's truly at stake?
- Are the other person pretending not to know important information about me or the situation involved? Would they have good reasons for doing so?
- Is there any reason you might pretend to be more naïve, ignorant, or incompetent in a particular area? Would you have good reasons for doing so?
- Is the other person's course of action too effective to be limited to what he shows to know or possess?
- Could the other person be pretending that the clear result of their strategy is actually due to external causes? By chance? Would they gain a clear advantage by doing so?
- What does the other person know (or is able to assume) about you and how might it want to manipulate, deceive, or mislead you by exploiting one of my weaknesses?

- What exactly does the other person expect you to do? Could they have anticipated that and already prepared a reaction that you could consider "unpredictable"?

Investigate the potential points of application for a "scientific hack"

- Can you take advantage of when the other is tired? Disoriented? Stressed?
- Can you exacerbate their condition of fatigue, stress, and confusion?
- You can take action during those moments, those phases, when their attention is diminished.
- Can you act in those times and moments when it would be completely unexpected?
- Can you take the other person to "play" in times, fields, and rules where they would be at a disadvantage?
- Can you take action in those moments, those phases when the other person's strength and energy are diminished?
- Can you take advantage of those moments when the other person is forced to open up or interact with something?
- Can you take advantage of the moments when the other person exhibits its maximum strength or power?
- Can you progressively take away the other person's resources, support, and foundations?

Alter someone else's table

- Does your behavior provide too many points of reference? Could this lead to a disadvantage for you? Could you act more "silently"?

- Can you lead the other person towards, or steer them away from, a specific direction by crafting a deception? Perhaps by lying about a particular topic?
- How could you lead the other person to believe something untrue?
- Can you craft something entirely fictional? Can you make it believable? Make it impossible to verify or disprove?
- Can you provide very little information and therefore lead the other person to instinctively fill his "gaps" with misleading conclusions?
- Can you plant clues that will automatically lead the other person to the truth we want him to construct?
- Can you slowly provide the other person with partial truths that lead they, step by step, to accept a larger truth?
- Can you confuse the other person by giving it too much information?
- Can you execute a "double bluff" and conspicuously pretend over something true? Bluff even when you have a "strong hand"?

Strategy or science?

I know what many might be thinking at this point in the book: all this strategic thinking and action might seem extremely far from the "laboratories" of science. However, in reality, these lines reveal even more the deep connection between science and strategy, firmly based on a "necessary" grasp of reality, as well as on continuously adapting to changing circumstances, lest they "lose" that grasp. In a sense, therefore, both disciplines are defined by being valuable tools for our species; universal "intellectual tools" with which we can attempt to tame and harness those "invisible forces" that govern everything. What is science, after all, if not the foremost strategy in the eternal battle of humanity against the darkest and most unstoppable forces of the universe?

Every invention, every building, every work of art we create is, for example, destined to deteriorate under the relentless weight of entropy; but science, with its indomitable determination, strives to counteract it, to slow it down, to conserve energy, to build more efficient and durable systems. But let us also consider diseases, those invisible forces that have threatened our species since the dawn of time. The plague, smallpox, tuberculosis: silent enemies that have decimated entire populations. And here comes the "strategic art" provided by science, which has allowed us to develop vaccines, therapies, and treatments; it has transformed certain deaths into preventable diseases and has granted humanity hope and life expectations never seen before. Consider also the most violent forces of nature: hurricanes, earthquakes, floods; calamities that have always had the power to even destroy entire civilizations. And even in this battle, science has always been by our side, providing us with the alphabets for the most advanced engineering, for sophisticated forecasting techniques, and for innovative intervention solutions.

But above all, science provides us with strategic tools to combat something more abstract and insidious: the drifts of superstition, ignorance, and dogmatic ideologies. Through its method, it offers us the best means to pave the way towards objectivity, clarity, and truth. In doing so, it reveals the guidelines for us to evolve, every day, towards the best we can represent as a species. The only major problem: this "path to betterment" is often far less appealing than falling into the easy answers offered by the alternative. Will we, therefore, be able to embrace the right path when it is asked of us?

Stories of "Scientific Champions"

Rita Levi-Montalcini: an Indomitable "Scientific Strategist"

The twentieth century bestowed upon science a veritable shower of geniuses, but among them, it can likely be said that an Italian light stands out, a woman with an indomitable spirit and a brilliant mind: Rita Levi-Montalcini.

Born in Turin in 1909, Rita grew up in a bourgeois family, surrounded by a society that favored domestic roles for women. But Rita was different. Driven by an insatiable passion for science, she decided to pursue a path in medicine, despite her father's objections.

The first true test for Rita, however, comes with the Second World War. Of Jewish faith, Rita and her family face racial discrimination enacted by the fascist regime. Instead of giving up, the woman, endowed with a tenacity that would define her entire career, sets up a clandestine laboratory in her bedroom in Turin. Here, amidst bombs and uncertainties, she begins to study the mystery of nerve tissue growth.

It is during these difficult years that she makes a discovery that will change her life and that of millions of other people. She identifies a protein that she calls "nerve growth factor" or NGF.

This protein plays a crucial role in the growth, survival, and maintenance of nerve cells.

Thanks to this discovery, the scientist gains recognition and begins collaborating with research institutes in America, where she continues her studies on NGF. Her discovery becomes crucial in the search for treatments for neurodegenerative diseases such as Alzheimer's and Parkinson's.

In 1986, her dedication and the significance of her research were finally acknowledged with the Nobel Prize in Medicine, which she shared with her colleague Stanley Cohen.

But Rita's story doesn't end here. Even after reaching the pinnacle of her scientific career, Rita doesn't stop. She founded the Institute of Neurobiology of the National Research Council in Rome and continued to work there until an advanced age.

More than any other aspect of her life, what clearly emerges from the story of Rita Levi-Montalcini is her unstoppable passion for science. Despite obstacles of all kinds, Rita never let herself be stopped. She always believed in the power of research, the beauty of science, and the ability of knowledge to change the world.

Her life, her challenges, and her discoveries are a true proof that perseverance, a love for research, and dedication can truly change the course of human history; and thus, her legacy still lives on in every researcher who, driven by the same passion, dedicates their existence to facing the unknown with the sole purpose of uncovering new, extraordinary truths for the well-being and progress of our species.

"Let's improve life through science and art."
(Virgil)

Thank you for reading!

I have always thought it would be incredibly difficult to discuss something as immense and extraordinary as science, and I'm also sure I've made mistakes or overlooked several aspects in this treatment of mine. And that's perfectly fine. Not only because admitting to being able to discuss such a monumental topic in a few pages would mean trivializing its significance. But also because the very fact that we live in a post-scientific revolution society gives us the ability to exchange information so quickly and in such detail that many of you will probably learn, study, and form new ideas precisely from everything I've gotten wrong or omitted. Which is simply magnificent.

Besides all the discussions about its success or lack thereof, my strongest hope is that, through this book, I have shared and perhaps strengthened within you some valuable principles. Perhaps reigniting your desire to embark on a stimulating intellectual journey. Rekindling a natural curiosity for study, research, and exploration. Helping you rediscover the thrill of surpassing the "Pillars of Hercules" of your conditioning and prejudices; all this, in a fascinating view of the unknown not as something fearful, but as a mystery to be unveiled, step by step.

As always, I wish you the best!

Danilo Lapegna

But there's more...

Official English website, with our full catalog:

https://kintsugiproject.net/pages/welcome

Instagram:

@danilolapegna.kintsugi

Do you have any feedback for us? Suggestions? Advice? Write to us at info@kintsugiproject.net

Because the material we have to offer certainly doesn't end with this book. In fact, we sincerely hope that this volume is just the beginning of a wonderful journey together!!

- The Science of Winning -

Danilo Lapegna

ULTRA HAPPINESS

A **COMPLETE, SCIENTIFIC GUIDE** FOR SPIRITUAL SURVIVAL IN THE MODERN WORLD.

The Kintsugi Project

- The Science of Winning -

Danilo Lapegna

THE SPEED MATH BIBLE

MASTER MENTAL MATH, BECOME A BETTER PROBLEM SOLVER, CONQUER YOUR DAILY CHALLENGES.

The Kintsugi Project

- The Science of Winning -

Danilo Lapegna

RESET!

HOW TO MASTER ART, SCIENCE AND PHILOSOPHY OF RADICAL, DURABLE CHANGE.

The Kintsugi Project

- The Science of Winning -

NEURO HACKING

Danilo Lapegna

NATURAL METHODS TO BOOST ENERGY, SLEEP DEEPLY, REDUCE STRESS AND LEARN FASTER.

The Kintsugi Project

Danilo Lapegna

THE CREATIVITY CODE

SCIENCE-BASED METHODS FOR GENERATING BRILLIANT IDEAS.

The Kintsugi Project

- The Science of Winning -

The Author

Danilo Lapegna, born in Italy in 1986, is the founder and CEO of the "Kintsugi Project." Based in Amsterdam, Netherlands, he is a tech project manager, a computer engineer, and an experienced writer with an insatiable passion for learning. From a young age, he showed an early fascination with the maximum potential of the human brain, devouring science-themed books and emerging as a television memory champion at the age of just six.

Thanks to his academic background in computer engineering, Danilo has successfully led international teams for years, working on high-impact software projects in the vibrant start-up scene of the United Kingdom. The complex management challenges within this highly competitive environment have fueled his growing passion and interest in a systemic and multidisciplinary approach to problems. This passion reaches its peak in his ability to generate value through rigorous data analysis and integration, driven by an unwavering desire to contribute to the well-being of others.

For over a decade, under the pseudonym "Yamada Takumi," he has leveraged his passions and skills to create a genuine "Scientific Quality Standard for Kintsugi," applied to books that have sold over 50,000 copies in Italy, climbing the sales charts on Amazon, helping thousands of people, and receiving enormous media attention for his success in the self-publishing industry.

And so, "The Kintsugi Project" represents the "ultimate" attempt by him and his staff to reinvent the approach to personal evolution, aiming to deconstruct all the "fluff" and the obsolete and dysfunctional paradigms of this sector. They then pivot their focus towards self-therapy systems, psychophysical well-being, "skill development," and "smart productivity" rooted in science, research, and above all, in a shared ecosystem that promotes individual and "personalized" growth, tailored to the values and needs of each person.

Bibliography and Further Reading

Book: "The Structure of Scientific Revolutions" by Thomas Kuhn (1962)

Book: "Game Theory: A Nontechnical Introduction" by Morton D. Davis (1970)

Book: "A Course in Probability Theory" by Kai Lai Chung (1974)

"Judgment under Uncertainty: Heuristics and Biases," by Tversky, Kahneman (1974)

Book: "The Selfish Gene," by Richard Dawkins (1976)

"The Framing of Decisions and the Psychology of Choice" by Tversky, Kahneman (1981)

"How the Laws of Physics Lie" by Cartwright (1983)

"Game Theory and Economic Modelling" by Kreps (1990)

"Measuring Individual Differences in Implicit Cognition: The Implicit Association Test" by Greenwald, McGhee, Schwartz (1998)

Book: "The Birth of Modern Science in Europe" by Paolo Rossi (2000)

Book: "The Black Swan: The Impact of the Highly Improbable" by Taleb (2007)

"The Pareto Managerial Principle: When Does It Apply?" by Grosfeld-Nir, Ronen, Kozlovsky (2007)

"Dragon-Kings, Black Swans and the Prediction of Crises" by Sornette (2009)

Book: "The Art of War" by Sun Tzu, Oscar Mondadori Edition (2011)

Book: "Thinking, Fast and Slow" by Kahneman (2013)

Book: "How Math Can Save Your Life" by James D. Stein (2013)

- The Science of Winning -

Disclaimer

Any information, reference, or advice related to the psychological, psychotherapeutic, biological, or medical sphere is not to be considered a substitute for any type of practice with a qualified professional. The reader is fully responsible for what they do with the data contained in this book and is obliged to consult healthcare professionals before making any decisions that could potentially impact their health.

For any passage contained herein and cited from other works, the author appeals to the right of citation and fair use regulations, and does not intend to economically harm any third-party authors or publishers in any way. All contents herein belong to their rightful owners, and the author encourages, where possible, support for the cited authors. For further information, please refer to the bibliography.

The scientific information included in this book is provided for educational and informational purposes. However, it is important to clarify that the authors have limited responsibilities regarding the accuracy, completeness, and timeliness of the information in the book. Science is an ever-evolving process, and some data may have already been replaced by new data or information. It is up to the reader to verify the sources and information in the book and act accordingly.

In general, it is recommended that readers always use their own judgment and refer to the most up-to-date and reliable sources of information before making any decisions based on the information contained in any section of this book.

The authors hold all commercial and non-commercial rights to the visual and informational material included in this book, in every form. Some images may have been generated using AI-based generative tools, but they have always been manually refined through human artistic work, preserving all copyright rights by the "Kintsugi Project."

www.ingramcontent.com/pod-product-compliance
Lightning Source LLC
Chambersburg PA
CBHW071531220526
45469CB00003B/727